T0222597

Advanced Structural Materials—2010

MATERIALS RESEARCH SOCIETY
SYMPOSIUM PROCEEDINGS VOLUME 1276

Advanced Structural Materials—2010

Symposium held August 15–19, 2010, Cancún, Mexico

EDITORS:

Hector A. Calderon

Instituto Politécnico Nacional
Mexico City, Mexico

Armando Salinas Rodriguez

Centro de Investigación y de Estudios Avanzados
del Instituto Politécnico Nacional Unidad Saltillo
Mexico City, Mexico

Heberto Balmori-Ramirez

Instituto Politécnico Nacional
Mexico City, Mexico

Materials Research Society
Warrendale, Pennsylvania

CAMBRIDGE UNIVERSITY PRESS
Cambridge, New York, Melbourne, Madrid, Cape Town,
Singapore, São Paulo, Delhi, Mexico City

Cambridge University Press
32 Avenue of the Americas, New York NY 10013-2473, USA

Published in the United States of America by Cambridge University Press, New York

www.cambridge.org
Information on this title: www.cambridge.org/9781107406766

Materials Research Society
506 Keystone Drive, Warrendale, PA 15086
http://www.mrs.org

© Materials Research Society 2010

First published 2010
First paperback edition 2012

Single article reprints from this publication are available through
University Microfilms Inc., 300 North Zeeb Road, Ann Arbor, MI 48106

CODEN: MRSPDH

ISBN 978-1-605-11253-4 Hardback
ISBN 978-1-107-40676-6 Paperback

CONTENTS

Preface ix

Materials Research Society Symposium Proceedings................... x

Oxidation Products in Inconel Alloys 600 and 690 Under
Hydrogenated Steam Environments and Their Role in Stress
Corrosion Cracking... 1
   Hugo F. Lopez

Mathematical Modeling of Impingement of an Air Jet
in a Liquid Bath.. 11
   J. Solórzano-López, R. Zenit,
   and M.A. Ramírez-Argáez

Manufacturing of Hybrid Composites and Novel Methods
to Synthesize Carbon Nanoparticles 21
   Andres E. Fals, Julio Quintero,
   and F.C. Robles Hernández

Effect of the Deposition Rate on Thin Films of CuZnAl
Obtained by Thermal Evaporation 31
   L. López-Pavón, E. López-Cuellar,
   A. Torres-Castro, C. Ballesteros,
   and C. José de Araújo

Development of an Algorithm for Random Packing
of Multi-Sized Spherical Particles 39
   H. de la Garza-Gutiérrez, G. Plascencia-Barrera,
   and S.D. de la Torre

Development of a New Nickel-Base Superalloy for High
Temperature Applications 49
   Octavio Covarrubias

Stress Ratio Effect on Fatigue Behavior of Aircraft Aluminum
Alloy 2024 T351 55
   M. Benachour, A. Hadjoui, M. Benguediab,
   and N. Benachour

v

Influence of Tempering Temperature in Wear of AISI T15 HSS Tools Produced by HIP and Liquid Phase Vacuum Sintering 61
Emmanuel P.R. Lima, Maurício D.M. das Neves,
Sérgio Delijaicov, and Francisco A. Filho

Effect of Hot Band Annealing on the Microstructure and Mechanical Properties of Low Carbon Electrical Steels 67
E. Gutiérrez-Castañeda and A. Salinas-Rodríguez

Microstructural and Mechanical Characterization of a TRIP-800 Steel Welded By Laser-CO_2 Process 73
G.Y. Perez-Medina, P. Zambrano, H.F. López,
F.A. Reyes-Valdés, and V.H. López-Cortés

Mechanism of Grain Growth During Annealing of Si-Al Electrical Steel Strips Deformed in Tension 79
J. Salinas B. and A. Salinas R.

Evolution of Microstructure of 304 Stainless Steel Joined by Brazing Process 85
F. García-Vázquez, I. Guzmán-Flores, A. Garza,
and J. Acevedo

Fourier Thermal Analysis of Eutectic Al-Si Alloy with Different Sr Content 91
R. Aparicio, G. Barrera, G. Trapaga,
and C. Gonzalez

Synthesis of Nanostructured Metal (Fe, Al)-C_{60} Composites 97
I.I. Santana García, V. Garibay Febles,
and H.A. Calderon

Metal-Graphite Couples Synthesized by Means of Mechanical Milling 103
I. Estrada-Guel, C. Carreño-Gallardo,
R. Pérez-Bustamante, J.M. Herrera-Ramírez,
and R. Martínez-Sánchez

Strengthening Phases in the Production of Al_{2024}-CNTs Composites by a Milling Process 111
R. Pérez-Bustamante, F. Pérez-Bustamante,
J.M. Herrera-Ramírez, I. Estrada-Guel,
P. Amézaga-Madrid, M. Miki-Yoshida,
and R. Martínez-Sánchez

**Wear Properties of an In-Situ Processed
TiC-Reinforced Bronze** 117
R. Sanchez and H.F. Lopez

**Nanoceria Coatings and Their Role on the High Temperature
Stability of 316L Stainless Steels** 123
H. Mendoza-Del-Angel and H.F. Lopez

**Preparation and Mechanical Characterization
of a Polymer-Matrix Composite Reinforced with PET** 129
J. Elena Salazar-Nieto, Alejandro Altamirano-Torres,
Francisco Sandoval-Pérez, and Enrique Rocha-Rangel

Micromechanical Models of Structural Behavior of Concrete 135
Ilya Avdeev, Konstantin Sobolev, Adil Amirjanov,
and Andrew Hastert

**Performance of Cement Systems with Nano- SiO₂ Particles
Produced Using Sol-Gel Method** 141
Konstantin Sobolev, Ismael Flores,
Leticia M. Torres, Enrique L. Cuellar,
Pedro L. Valdez, and Elvira Zarazua

**Preparation and Mechanical Characterization of Composite
Material of Polymer-Matrix with Starch Reinforced
with Coconut Fibers** 149
Yaret G. Torres-Hernández,
Alejandro Altamirano-Torres,
Francisco Sandoval-Pérez,
and Enrique Rocha Rangel

**Magnetic Properties of Bulk Composite FeBSiM (M=Cr,Zr)
Alloys with High Microhardness** 155
S. Báez, I. Betancourt, M.E. Hernandez-Rojas,
and I.A. Figueroa

**Microstructure Characterization of Textured Nickel Using
Parameters of Extinction.** 163
A. Cadena Arenas, T. Kryshtab, J. Palacios Gómez,
G. Gómez Gasga, A. De Ita de la Torre,
and A. Kryvko

**Mechanofusion Processing of Metal-Oxide Composite Powders
for Plasma Spraying** 169
Ricardo Cuenca-Alvarez,
Carmen Monterrubio-Badillo, Hélène Ageorges,
and Pierre Fauchais

Stress Concentration on Artificial Pitting Holes and Fatigue Life for Aluminum Alloy 6061-T6, Undergoing Rotating Bending Fatigue Tests 175
Víctor H.M. Lemus, Gonzalo M.D. Almaraz,
and J. Jesús V. Lopez

Synthesis of $La_4Ni_3O_{10}$ Cathode Material (SOFC) by Sol-Gel Process 181
Rene Fabian Cienfuegos, Sugeheidy Carranza,
Leonardo Chávez, Laurie Jouanin, Guillaume Marie,
and Moisés Hinojosa

Alumina-Based Functional Materials Hardened with Al or Ti and Al-Nitride or Ti-Nitride Dispersions 187
José G. Miranda-Hernández,
Elizabeth Refugio-Garcia, Elizabeth Garfias-García,
and Enrique Rocha-Rangel

Author Index 193

Subject Index 195

PREFACE

Symposium 4, "Advanced Structural Materials," held August 15–19, 2010, in Cancún, Mexico, was part of the XIX International Congress on Materials Research organized jointly by the Mexican Materials Research Society (MRS Mexico) and the Materials Research Society (MRS). The symposium was devoted to fundamental and technological applications of structural materials, and continued the tradition of providing a forum for scientists from various backgrounds with a common interest in the development and use of structural materials to come together and share their findings and expertise.

The papers contained in this volume are a collection of invited and contributed papers. This year, the symposium was attended by participants from France, Japan, Mexico, Turkey, the United States and Spain. We are grateful to all of the referees who, by their comments and constructive criticism, helped to improve the final printed papers, and to all the authors who made additional efforts to prepare their manuscripts.

This symposium was the latest in a series held over the last 13 years with the objective of presenting an overview and the most recent investigations into advanced structural metallic, ceramic and composite materials. The topics include innovative processing, phase transformations, mechanical properties, oxidation resistance, modeling and the relationship between processing, microstructure, and mechanical behavior. Additionally, papers on industrial application and metrology are included.

The organizing committee gratefully acknowledges the enthusiastic cooperation of all symposium participants, as well as the kind acceptance of the Materials Research Society staff to publish these proceedings. The financial support of the Instituto Politécnico Nacional (Mexico) and the Centro de Investigación y de Estudios Avanzados del Instituto Politécnico Nacional is also acknowledged.

Hector A. Calderon
Armando Salinas Rodríguez
Heberto Balmori-Ramírez

November 2010

MATERIALS RESEARCH SOCIETY SYMPOSIUM PROCEEDINGS

Volume 1245 — Amorphous and Polycrystalline Thin-Film Silicon Science and Technology—2010, Q. Wang, B. Yan, C.C. Tsai, S. Higashi, A. Flewitt, 2010, ISBN 978-1-60511-222-0

Volume 1246 — Silicon Carbide 2010—Materials, Processing and Devices, S.E. Saddow, E.K. Sanchez, F. Zhao, M. Dudley, 2010, ISBN 978-1-60511-223-7

Volume 1247E —Solution Processing of Inorganic and Hybrid Materials for Electronics and Photonics, 2010, ISBN 978-1-60511-224-4

Volume 1248E —Plasmonic Materials and Metamaterials, J.A. Dionne, L.A. Sweatlock, G. Shvets, L.P. Lee, 2010, ISBN 978-1-60511-225-1

Volume 1249 — Advanced Interconnects and Chemical Mechanical Planarization for Micro- and Nanoelectronics, J.W. Bartha, C.L. Borst, D. DeNardis, H. Kim, A. Naeemi, A. Nelson, S.S. Papa Rao, H.W. Ro, D. Toma, 2010, ISBN 978-1-60511-226-8

Volume 1250 — Materials and Physics for Nonvolatile Memories II, C. Bonafos, Y. Fujisaki, P. Dimitrakis, E. Tokumitsu, 2010, ISBN 978-1-60511-227-5

Volume 1251E —Phase-Change Materials for Memory and Reconfigurable Electronics Applications, P. Fons, K. Campbell, B. Cheong, S. Raoux, M. Wuttig, 2010, ISBN 978-1-60511-228-2

Volume 1252 — Materials and Devices for End-of-Roadmap and Beyond CMOS Scaling, A.C. Kummel, P. Majhi, I. Thayne, H. Watanabe, S. Ramanathan, S. Guha, J. Mannhart, 2010, ISBN 978-1-60511-229-9

Volume 1253 — Functional Materials and Nanostructures for Chemical and Biochemical Sensing, E. Comini, P. Gouma, G. Malliaras, L. Torsi, 2010, ISBN 978-1-60511-230-5

Volume 1254E —Recent Advances and New Discoveries in High-Temperature Superconductivity, S.H. Wee, V. Selvamanickam, Q. Jia, H. Hosono, H-H. Wen, 2010, ISBN 978-1-60511-231-2

Volume 1255E —Structure-Function Relations at Perovskite Surfaces and Interfaces, A.P. Baddorf, U. Diebold, D. Hesse, A. Rappe, N. Shibata, 2010, ISBN 978-1-60511-232-9

Volume 1256E —Functional Oxide Nanostructures and Heterostructures, 2010, ISBN 978-1-60511-233-6

Volume 1257 — Multifunctional Nanoparticle Systems—Coupled Behavior and Applications, Y. Bao, A.M. Dattelbaum, J.B. Tracy, Y. Yin, 2010, ISBN 978-1-60511-234-3

Volume 1258 — Low-Dimensional Functional Nanostructures—Fabrication, Characterization and Applications, H. Riel, W. Lee, M. Zacharias, M. McAlpine, T. Mayer , H. Fan, M. Knez, S. Wong, 2010, ISBN 978-1-60511-235-0

Volume 1259E —Graphene Materials and Devices, M. Chhowalla, 2010, ISBN 978-1-60511-236-7

Volume 1260 — Photovoltaics and Optoelectronics from Nanoparticles, M. Winterer, W.L. Gladfelter, D.R. Gamelin, S. Oda, 2010, ISBN 978-1-60511-237-4

Volume 1261E —Scanning Probe Microscopy—Frontiers in NanoBio Science, C. Durkan, 2010, ISBN 978-1-60511-238-1

Volume 1262 — In-Situ and Operando Probing of Energy Materials at Multiscale Down to Single Atomic Column—The Power of X-Rays, Neutrons and Electron Microscopy, C.M. Wang, N. de Jonge, R.E. Dunin-Borkowski, A. Braun, J-H. Guo, H. Schober, R.E. Winans, 2010, ISBN 978-1-60511-239-8

Volume 1263E —Computational Approaches to Materials for Energy, K. Kim, M. van Shilfgaarde, V. Ozolins, G. Ceder, V. Tomar, 2010, ISBN 978-1-60511-240-4

Volume 1264 — Basic Actinide Science and Materials for Nuclear Applications, J.K. Gibson, S.K. McCall, E.D. Bauer, L. Soderholm, T. Fanghaenel, R. Devanathan, A. Misra, C. Trautmann, B.D. Wirth, 2010, ISBN 978-1-60511-241-1

Volume 1265 — Scientific Basis for Nuclear Waste Management XXXIV, K.L. Smith, S. Kroeker, B. Uberuaga, K.R. Whittle, 2010, ISBN 978-1-60511-242-8

Volume 1266E —Solid-State Batteries, S-H. Lee, A. Hayashi, N. Dudney, K. Takada, 2010, ISBN 978-1-60511-243-5

Volume 1267 — Thermoelectric Materials 2010—Growth, Properties, Novel Characterization Methods and Applications, H.L. Tuller, J.D. Baniecki, G.J. Snyder, J.A. Malen, 2010, ISBN 978-1-60511-244-2

Volume 1268 — Defects in Inorganic Photovoltaic Materials, D. Friedman, M. Stavola, W. Walukiewicz, S. Zhang, 2010, ISBN 978-1-60511-245-9

Volume 1269E —Polymer Materials and Membranes for Energy Devices, A.M. Herring, J.B. Kerr, S.J. Hamrock, T.A. Zawodzinski, 2010, ISBN 978-1-60511-246-6

MATERIALS RESEARCH SOCIETY SYMPOSIUM PROCEEDINGS

Volume 1270 — Organic Photovoltaics and Related Electronics—From Excitons to Devices, V.R. Bommisetty, N.S. Sariciftci, K. Narayan, G. Rumbles, P. Peumans, J. van de Lagemaat, G. Dennler, S.E. Shaheen, 2010, ISBN 978-1-60511-247-3

Volume 1271E —Stretchable Electronics and Conformal Biointerfaces, S.P. Lacour, S. Bauer, J. Rogers, B. Morrison, 2010, ISBN 978-1-60511-248-0

Volume 1272 — Integrated Miniaturized Materials—From Self-Assembly to Device Integration, C.J. Martinez, J. Cabral, A. Fernandez-Nieves, S. Grego, A. Goyal, Q. Lin, J.J. Urban, J.J. Watkins, A. Saiani, R. Callens, J.H. Collier, A. Donald, W. Murphy, D.H. Gracias, B.A. Grzybowski, P.W.K. Rothemund, O.G. Schmidt, R.R. Naik, P.B. Messersmith, M.M. Stevens, R.V. Ulijn, 2010, ISBN 978-1-60511-249-7

Volume 1273E —Evaporative Self Assembly of Polymers, Nanoparticles and DNA , B.A. Korgel, 2010, ISBN 978-1-60511-250-3

Volume 1274 — Biological Materials and Structures in Physiologically Extreme Conditions and Disease, M.J. Buehler, D. Kaplan, C.T. Lim, J. Spatz, 2010, ISBN 978-1-60511-251-0

Volume 1275 — Structural and Chemical Characterization of Metals, Alloys and Compounds, R. Pérez Campos, A. Contreras Cuevas, R.A. Esparza Muñoz, 2010, ISBN 978-1-60511-252-7

Volume 1276 — Advanced Structural Materials—2010, H.A. Calderon, A. Salinas-Rodríguez, H. Balmori-Ramírez, 2010, ISBN 978-1-60511-253-4

Volume 1277E —Biomaterials—2010, S.E. Rodil, A. Almaguer-Flores, K. Anselme, 2010, ISBN 978-1-60511-254-1

Volume 1278E — Composite, Hybrid Materials and Ecomaterials, R. Bernal, C. Cruz, L. Rendon, V.M. Castano Best, 2010, ISBN 978-1-60511-255-8

Volume 1279 — New Catalytic Materials, J-A. Wang, J. Manuel Domínguez , 2010, ISBN 978-1-60511-256-5

Volume 1280E — Nanomaterials for Biomedical Applications, L. Zhang, A. Salinas-Rodríguez, and T.J. Webster, 2010, ISBN 978-1-60511-257-2

Prior Materials Research Society Symposium Proceedings available by contacting Materials Research Society

Mater. Res. Soc. Symp. Proc. Vol. 1276 © 2010 Materials Research Society

Oxidation Products in Inconel Alloys 600 and 690 Under Hydrogenated Steam Environments and Their Role in Stress Corrosion Cracking.

Hugo F. Lopez
Materials Department, University of Wisconsin-Milwaukee
3200 N. Cramer St. Milwaukee WI 53209,

ABSTRACT

Thermodynamic considerations for the stability of Ni and Cr compounds developed under PWR environments (P_{H2O} and P_{H2}) are experimentally tested. In particular, the experimental outcome indicates that $Ni(OH)_2$ and $CrOOH$ are thermodynamically stable products under actual PWR conditions (T > 360°C and Pressures of up to 20 MPa). Accordingly, a mechanism is proposed to explain crack initiation and growth in inconel alloy 600 along the gbs. The mechanism is based on the existing thermodynamic potential for the transformation of a protective NiO surface layer into an amorphous non-protective $Ni(OH)_2$ gel. This gel is also expected to form along the gbs by exposing the gb Ni-rich regions to H_2 supersaturated water steam. Crack initiation is then favored by tensile stressing of the gb regions which can easily rupture the brittle gel film. Repeating the sequence of reactions as fresh Ni is exposed to the environment is expected to also account for crack growth in Inconel alloy 600. The proposed crack initiation mechanism is not expected to occur in alloy 690 where a protective Cr_2O_3 film covers the metal surface. Yet, if a pre-existing crack is present in alloy 690, crack propagation would occur in the same manner as in alloy 600.

INTRODUCTION

Inconel Alloys 600 used in pressurized water reactor (PWR) environments are often found to undergo stress corrosion cracking (SCC) [1-3]. Under these conditions, hydrogen supersaturated steam at temperatures above 300°C in combination with a stressed susceptible microstructure can lead to crack development along the grain boundaries (gbs). It has been observed that cracks always initiate at grain boundaries, but they will not initiate in single crystals [4]. The susceptibility of alloy 600 to undergo SCC has lead to costly repairs and to seek other alloy alternatives such as Inconel alloy 690. The main difference between these two alloys is in the Cr content (15 wt% in alloy 600 vs. 30 wt% in alloy 690).

From the published literature, it is apparent that alloy 690 is not susceptible of undergoing crack initiation [5,6]. Yet, when a crack has been initiated by some other means such as fatigue cracking, it will be able to propagate at rates similar to those exhibited in alloy 600 [7,8]. In turn, this strongly suggests that the Cr content might be able to inhibit crack initiation, yet it has little effect on crack propagation. In addition, current research indicates that alloy 600 will not crack in pure steam or in dry H_2 environments [9, 10]. Hence, any proposed cracking mechanism must account for the role of H_2 in combination with water steam. In particular, it has been reported that the highest SCC susceptibility occurs when the H_2 partial pressures are near the Ni/NiO stability lines [11]. In this work, thermodynamic calculations are carried out and stability diagrams are constructed for Ni and Cr compounds under PWR environments (P_{H_2O} and P_{H_2}).

From these determinations a mechanism is proposed that is able to account for IGSCC in Inconel alloy 600.

THEORY

Thermodynamic Analysis

Table 1 gives the thermodynamic Gibbs free energies for various potential reactions between Ni or Cr with water steam and hydrogen. Various reaction products are considered in this table including oxides, hydroxides and oxyhydroxides as they have been experimentally reported in Ni-Cr alloys [12-15]. In the case of $Ni(OH)_2$, experimental as well as theoretical predictions for the Gibbs free energy of formation are available, but data is lacking for CrOOH. Hence, some theoretical estimates are made in this work.

Table 1. Reaction Thermodynamic Data for Ni and Cr

Item	Compound or Reaction	298 K			600 K			Ref	Notes
		ΔH / (J/mol)	ΔS_1 / (J/mol*K)	ΔG°_{298} / (J/mol)	ΔH / (J/mol)	ΔS_1 / (J/mol*K)	ΔG°_{600} / (J/mol)		
1	$O_{2(g)}$	0	205.146	-61,134	9,303	226.572	-126,640	27	
2	$H_{2(g)}$	0	130.000	-38,740	8,776	151.004	-81,826	27	
3	$H_2O_{(g)}$	-241,856	188.824	-298,126	-231,130	213.441	-359,195	27	
4	Cr	0	23.640	-7,045	7,749	41.446	-17,119	27	
5	Cr_2O_3	-1,140,558	81.170	-1,164,747	-1,105,365	162.517	-1,202,875	27	
6	CrOOH			-732,373			-773,938		Assumed 1st hydroxylation is (ΔG_{oxide} -300,000)/2 at 298 K and (ΔG_{oxide} -345000)/2 at 600 K
7	$Cr(OH)_3$			-1,000,373			-1,086,938		Assumed 2nd hydroxylation is ΔG_{oxide} -300,000+32,000 at 298 K and ΔG_{oxide} -300,000+32,000 at 600 K
8	Ni	0	29.874	-8,902	8,988	50.420	-21,264	27	
9	NiO	-239,743	37.991	-251,064	-225,059	72.451	-268,530	27	
10	$Ni(OH)_2$	-529,700	88.000	-555,924			-608,530	21	Hydroxylation is approx -305,000 at 298 K. Assumed hydroxylation energy is ΔG_{oxide} - 340,000 at 600K
11	2 Cr + 3 H_2O = Cr_2O_3 + 3 H_2	-414,990	-142.582	-372,501	-401,145	-107.686	-336,533		
12	Cr + 2 H_2O = CrOOH + 3/2 H_2			-187,188			-161,169		Used estimated Data for CrOOH
13	Cr + 3 H_2O = $Cr(OH)_3$ + 3/2 H_2			-157,062			-114,975		Used estimated Data for $Cr(OH)_3$
14	Cr_2O_3 + H_2O = 2 CrOOH			-1,874			14,195		Used estimated Data for CrOOH
15	CrOOH + H_2O = $Cr(OH)_3$			30,126			46,195		Used estimated Data for CrOOH and $Cr(OH)_3$
16	Ni + H_2O = NiO + H_2	2,113	-50.707	17,224	5,859	-40.406	30,103		
17	Ni + 2 H_2O = $Ni(OH)_2$ + H_2	-45,988	-189.522	10,490			49,297		Used estimated Data for $Ni(OH)_2$
18	NiO + H_2O = $Ni(OH)_2$	-48,101	-138.815	-6,734			19,195		Used estimated Data for $Ni(OH)_2$

2

From the thermodynamic data, phase stability diagrams are constructed for the Ni and Cr systems at 600 K as a function of the H_2O and H_2 pressures (see Fig.1). Accordingly, it is evident from thermodynamic arguments that the precipitation of an oxide or hydroxide product will depend on the actual partial hydrogen pressures and partial steam pressures. In addition, a critical steam pressure exists below which NiO is the stable phase.

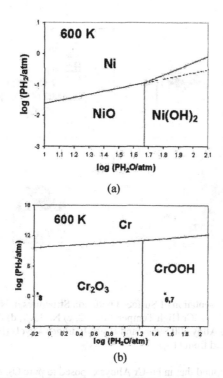

(a)

(b)

Figure 1. (a) Ni/NiO/Ni(OH)$_2$ Stability Diagram at 600 K and (b) Cr/Cr$_2$O$_3$/CrOOH Stability Diagram at 600 K

RESULTS AND DISCUSSION

Oxidation of Ni-Cr Alloys

Since crack initiation is a surface phenomenon, alloy oxidation either in pure O_2, pure H_2O, or H_2O/H_2 mixtures is expected to provide key information on the active cracking mechanisms. Figure 1 is a schematic representation of the developed oxide layers reported [12, 13] under each of the aforementioned environments.

3

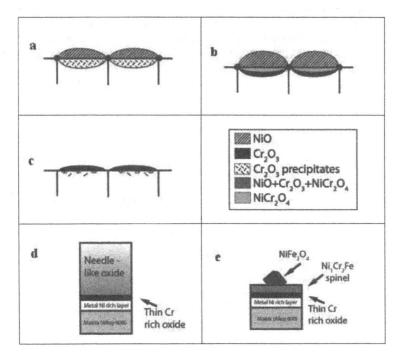

Figure 2. Schematic Representation of Surface Oxidation Structures in Ni-Cr Alloys a) Ni-20Cr Low Temperature O2, b) Ni-20Cr High Temperature O2, c) Ni-33Cr, d) Alloy 600 in Steam and Low Hydrogen Levels, e) Alloy 600 Steam and High Hydrogen Levels (figures a-c adapted from [12] figures d and e adapted from [13])

In particular, it has been found that in Ni-Cr Alloys exposed to pure O$_2$, the type and structure of the surface oxide layers depends on both, the Cr concentration and the exposure temperature [12-15]. In Alloys containing 10-30 wt% Cr (i.e. those analogous to alloy 600) NiO and Cr$_2$O$_3$ form simultaneously on the alloy surface, but NiO growth kinetics is prevalent and NiO rapidly develops a continuous layer. Isolated Cr$_2$O$_3$ islands are found to react with the NiO layer giving rise to NiCr$_2$O$_4$ spinels [13].

In addition, Cr$_2$O$_3$ forms preferentially at gbs where the Cr diffusion rate is relatively high. Hence, given time a Cr$_2$O$_3$ sublayer will grow outwards from underneath the NiO layer by internal oxidation. As long as Cr diffusion is limited due to the relatively low temperatures (below 600°C), the isolated gb Cr$_2$O$_3$ film will not be able to form a continuous scale (Figure 2a). Nevertheless, at increasing temperatures, Cr diffusion rates are high enough to promote the development of a continuous sublayer through lateral Cr$_2$O$_3$ growth along the gbs. In turn this

4

will result in a Cr depleted region beneath the Cr_2O_3 sublayer (Figure 2b). The preferred growth direction of Cr_2O_3 would be towards the Cr rich regions. Hence, it is reasonable to expect that the metal next to the Cr_2O_3 layer would be depleted in Cr. Oxidation studies on alloy 600 in pure steam at PWR operating temperatures show a 3-layered structure reported for low temperatures in Ni-20Cr Alloys [12]. In particular, when the environment is pure O_2 columnar oxide scales are formed, but when the alloy is under water vapor a needle-like scale is produced.

In alloys containing above 30 wt% Cr (i.e. those analogous to Alloy 690), NiO and Cr_2O_3 still form simultaneously on the exposed surfaces. However, in this case, the high Cr levels apparently promote Cr_2O_3 growth kinetics to dominate resulting in a continuous Cr_2O_3 layer. Moreover, isolated NiO islands react with the Cr_2O_3 layer to form $NiCr_2O_4$. (Figure 2c). Introduction of H_2 into H_2O steam is found to modify the surface oxide layered structure depending on the hydrogen partial pressures [12]. Accordingly, for H_2 pressures under which NiO is stable a four-layered structure similar to that found in Ni-20Cr alloys is formed (Figure 2d). Yet, H_2 pressures under which only metallic Ni is stable lead to a NiO free surface layer and a 3-layered structure consisting of a thin $NiCr_2O_4$ layer, a continuous Cr_2O_3 layer, and a Cr depleted layer (Figure 2e).

As stated before, in PWR environments, inconel alloys 600 are exposed to hydrogen pressures near the Ni/NiO equilibrium boundary lines. Under these conditions, the alloy is most susceptible to exhibit crack initiation. Accordingly, the exhibited oxide morphology within cracks and along the gbs generally conforms to the 4-layered oxide structure of the high temperature Ni-20Cr alloys (Figure 2f) [14]. In addition, the development of precursor hydroxides is thermodynamically viable under these environments. However, it is difficult to clearly resolve the presence of hydroxides as they are expected to decompose into the actual oxides when removed from the PWR environments [14]. Evidence for the development of hydroxides has been reported [11] suggesting that in H_2 saturated water containing LiOH (pH=7) at 285°C, all outer layers are MeOOH or $Me(OH)_2$ hydroxide species.

Stability Diagrams

Under the PWR environments and based on the predictions of the P_{H2O}-P_{H2} stability diagrams (Fig. 1) it is evident that the surface oxidation products should be $Ni(OH)_2$ and CrOOH. Both of these phases can be either crystalline or amorphous gels capable of absorbing appreciable amounts of water. Hence, they do not constitute a protective oxidation barrier as water (absorbed in the hydroxide gel) is always in intimate contact with the bare metal. Yet, these corrosion products are not at first present on the metal surfaces. Initially, a NiO protective layer covers the alloy 600 surface, with Cr_2O_3 expected to be present along the gbs intersecting the surface. In alloy 690, Cr_2O_3 is expected to be the protective oxide layer, and it might also be present along the gbs.

Hence, in order to initiate a crack, the protective surface oxides must be transformed into non-protective hydroxides. In the case of NiO, Aia et al [16] found that it is not likely to hydrate crystalline NiO (i.e. convert NiO to $Ni(OH)_2$) with only water. Apparently, this is due to the lack of actual H_2O molecules in the hydroxide lattice and typically conversion to NiO occurs by dehydroxylation. The hydration of NiO under PWR conditions can occur as hydrogen is present in these environments. As already stated, the highest susceptibility for crack initiation in alloy 600 occurs when the P_{H2} approaches the Ni/NiO stability boundary.

From the Ni stability diagram (Figure 1a) it can be seen that $Ni(OH)_2$ is stable at P_{H2} above the Ni/NiO boundary line. Accordingly, at a P_{H2} just above the Ni/NiO boundary, NiO would be reduced to Ni by the reverse of reaction 1 (i.e. NiO + H_2 => Ni + H_2O). However, $Ni(OH)_2$ is still stable in this region. Therefore, metallic nickel from the NiO reduction would be free to react with water according to:

$$Ni + 2\ H_2O\ =\ Ni(OH)_2 + H_2 \qquad (1)$$

The overall reaction is then:

$$NiO + H_2O = Ni(OH)_2 \qquad (2)$$

Thus, it is thermodynamically possible that NiO can transform into $Ni(OH)_2$ at a P_{H2} near the Ni/NiO boundary. Moreover, the rate of conversion would increase with increasing P_{H2} up to a maximum corresponding to the metal/oxide equilibrium condition.

Assuming that similar mechanisms are required for the transformation of Cr_2O_3 into CrOOH, it can be concluded that under the actual PWR environments the P_{H2O}-P_{H2} conditions are far below the Cr/Cr_2O_3 equilibrium lines. Hence, it can be inferred that only alloys with predominantly NiO surface oxides would be susceptible to undergo crack initiation. This agrees with published reports of crack initiation in alloy 600 where the surface oxide is NiO [1-3], and of the apparent stability of alloy 690 which is expected to be protected by a Cr_2O_3 layer [5,6]. Hence, the effect of hydrogen on the crack susceptibility of inconel 600 under PWR environments can be accounted for. Further evidence for the effect of hydrogen is given U. Ehrnsten [17] who measured electrical conductivities on the surfaces of a growing crack in an inconel alloy 600 under simulated PWR environments. In their work, they reported a drop in electrical resistance as hydrogen is increased in the PWR environment. In turn, this effect can be satisfactorily explained by the conversion of NiO to a $Ni(OH)_2$ gel containing water, a material with better conductivity than NiO. Since the increase in electrical conductivity is found only when the oxide is exposed to H_2 in the steam, this is in line with the proposed arguments.

Proposed IGSCC Mechanism

Figure 3 shows schematically a possible sequence of crack initiation and propagation steps in alloy 600. Initially, the metal surface is clean and without an oxide layer (Figure 3a). After exposure to O_2, NiO has formed on the alloy surface and Cr_2O_3 is present along the gbs. Beneath the Cr_2O_3 developed at the gbs a Ni-rich region is expected to be present (Figure 3b). When the metal is exposed to supersaturated steam at a P_{H2O} above the critical pressure thermodynamic conditions favor the NiO to $Ni(OH)_2$ transformation (Figure 3c). Since absorbed water in the gel comes into intimate contact with the bare metal alloy, the Ni-rich regions along the gbs react to form $Ni(OH)_2$ while the Cr-rich regions give rise to CrOOH. In addition, the Cr_2O_3 which originally formed along the gbs becomes isolated and reacts with the surrounding NiO or $Ni(OH)_2$ to form spinel oxides (Figure 3d). These oxidation reactions are expected to occur over

6

the entire alloy surface, as well as at the gbs intersecting the surface and give rise to the reported 4-layered oxide structures [12, 13].

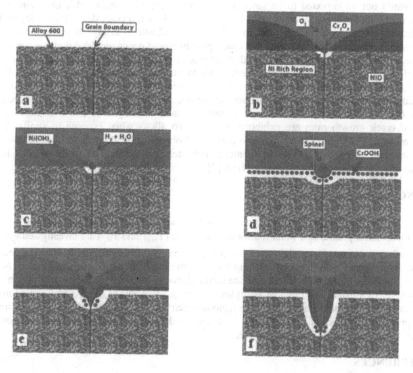

Figure 3. Proposed Crack Initiation and Propagation Mechanism in Alloy 600.

From TEM studies [14], the active gb oxidation reactions have been described as intergranular attack (IGA). In addition, spinel oxides have been reported to form frequently either at the crack entry, or if wide enough, inside cracks [12, 13]. This agrees with the mechanism proposed in this work as the gb Cr_2O_3 acts as a precursor in promoting a Ni-rich region which can eventually lead to crack formation directly beneath the gbs. Moreover, in gb regions where tensile stresses are present, it is likely that they will induce crack initiation by rupturing the brittle $Ni(OH)_2$ film (Figure 3e). As a result, exposure of fresh Ni metal to the PWR environment arises repeating the sequence of oxidation reactions previously described. Then, when the hydroxide developed at the gbs can no longer support the applied load, it will fracture leading to further crack propagation (Figure 3f).

In the absence of a gb structure, Cr_2O_3 and the accompanying Ni-rich region would not be able to develop and cracks would not form. This explains why in alloy 600 single crystals no crack initiation is observed [4]. Moreover, in alloy 690 a protective Cr_2O_3 layer covers the entire

7

surface, and no NiO is available that can be converted into $Ni(OH)_2$, thus avoiding any intimate contact between the H_2 supersaturated water steam and the underlying surface. Therefore, alloy 690 would not be expected to be susceptible to crack initiation under PWR environments in agreement with the experimental reports [5, 6]. Nevertheless, if a crack develops in alloy 690 by mechanical stressing including fatiguing, then crack growth is always possible. From the proposed mechanism, it is expected that crack growth would occur provided that any protective oxide at the surface of the crack tip either insufficiently covers the tip, or cracks under local stressing. Accordingly, exposure of fresh metal to the PWR environment could result in the development of non-protective hydroxide gels at the crack tip as predicted from thermodynamic considerations. In turn, crack propagation would occur by the same mechanism proposed for alloy 600. Preliminary work on crack propagation rates in alloys 600 and 690 indicate that similar crack growth rates are exhibited in these two alloys when exposed to similar PWR environments [7,8]. Moreover, high resolution TEM observations of crack surfaces in thin foils of alloy 690 exposed to PWR environments indicates the presence of an amorphous layer, suggesting the presence of a hydroxide phase [7].

CONCLUSIONS

In this work thermodynamic arguments for the stability of Ni and Cr compounds developed under pressurized water reactor environments (P_{H_2O} and P_{H_2}) are investigated. A mechanism is proposed to explain crack initiation and propagation in alloy 600 along the grain boundaries where Cr_2O_3 forms by leaching out Cr from the matrix leaving behind a porous Ni-rich region. The mechanism is based on the thermodynamic potential for the transformation of a protective NiO surface layer into an amorphous non-protective $Ni(OH)_2$ gel. Crack initiation is then favored by tensile stressing of the gb regions which can easily rupture the gelatinous film. The proposed crack initiation mechanism is not expected to occur in alloy 690 where a protective Cr_2O_3 film covers the entire metal surface.

REFERENCES

1. N. Totuska, E. Lunarska, G. Grangnolino, and Z. Szklarska-Smialowska, Corrosion, Vol. 43, 1987, pp. 505-514.
2. R. B. Rebak and Z. Szklarska-Smialowska, Corrosion, Vol. 47, 1991, pp. 754-757.
3. Z. Szklarska-Smialowska, W-K. Lai, Z. Xia, Corrosion. Vol. 46, No. 10. 1990. pp. 853-860.
4. T. Magnin, N. Renaudot, F. Foct, Materials Transactions, JIM, Vol. 41, No. 1, 2000 pp. 210-218.
5. J.-D. Mithieux, et al., Eurocorr '96, IV/IX, 1996, pp. 2-1 - 2-4.
6. C. M. Brown and W. J. Mills, Corrosion, Vol. 55, No. 2, 1999, pp. 173-185.
7. M. Ali, "Environmentally Induced Crack Propagation in Inconel Alloys 600 and 690 Under Supersaturated Steam," Ph.D. Thesis, University of Wisconsin - Milwaukee, 2004.
8. H. F. Lopez and J. B. Ferguson, SIF2004, Conf. Proc, eds. A. Atrens et al., Sept. 2004, pp. 249-256.
9. C. Shen, P. Shewmon, Metallurgical Transactions A, Vol. 21A, 1990 pp. 1261-1271.
10. R. Rios, T. Magnin, D. Noel, O. de Bouvier, Metallurgical and Materials Transactions A, Vol. 26A, 1995, pp. 925-939.

11. P. Scott, Corrosion, Vol. 56, No. 8, 2000, pp. 771-782.
12. G. Calvarin, R. Molins, A. Huntz, Oxidation of Metals, Vol. 53, Nos. 1/2, 2000, pp. 25-48.
13. T. Terachi, N. Totsuka, T. Yamada, T. Nakagawa, H. Deguchi, M. Horiuchi, M. Oshitani, Journal of Nuclear Science and Technology, Vol. 40, No. 7, 2003, pp. 509-516.
14. G. Calvarin, R. Molins, A. Huntz, Oxidation of Metals, Vol. 54, Nos. 1/2, 2000, pp. 399-426.
15. S. Lozano-Perez, J. Titchmarsh, Materials at High Temperatures, 20(4), 2003, pp. 573-579.
16. M. Aia, Journal of the Electrochemical Society, Vol. 113, No. 10, 1966, pp. 1045-1047.
17. U. Ehrnsten, et al., Eurocorr '96 IV/IX, 1996, pp. 6-1 - 6-4.

Mater. Res. Soc. Symp. Proc. Vol. 1276 © 2010 Materials Research Society

Mathematical Modeling of Impingement of an Air Jet in a Liquid Bath.

J. Solórzano-López[1], R. Zenit[2], M.A. Ramírez-Argáez[1].

[1] Facultad de Química, Universidad Nacional Autónoma de México, México, D.F.
[2] Instituto de Investigaciones en Materiales, Universidad Nacional Autónoma de México.
xaxni2006@yahoo.com.mx, zenit@unam.mx, marco.ramirez@servidor.unam.mx.

ABSTRACT

Physical and mathematical modeling of jet-bath interactions in electric arc furnaces represent valuable tools to obtain a better fundamental understanding of oxygen gas injection into the furnace. In this work, a 3D mathematical model is developed based on the two phase approach called Volume of Fluid (VOF), which is able to predict free surface deformations and it is coded in the commercial fluid dynamics software FLUENT™. Validation of the mathematical model is achieved by measurements on a transparent water physical model. Measurements of free surface depressions through a high velocity camera and velocity patterns are recorded through a Particle Image Velocimetry (PIV) Technique. Flow patterns and depression geometry are identified and characterized as function of process parameters like distance from nozzle to bath, gas flow rate and impingement angle of the gas jet into the bath. A reasonable agreement is found between simulated and experimental results.

INTRODUCTION

The oxygen jets are used in several steelmaking processes such the Electric Arc Furnace (EAF). These jets play an important role in these processes because they control bath mixing, chemical reaction kinetics, energy consumption, foaming slag formation, bath recirculation and the occurrence of splashing since they exchange momentum, heat and mass with the slag and the molten steel [1, 2]. To evaluate the momentum exchange between the gas jets and the liquid bath, the geometry of the formed depression in the liquid free surface must be determined as well as the liquid velocity profiles induced by the impingement and shear of the gas jet respectively.

Mathematical modeling is a tool able to simulate these complex systems. Also, physical modeling, using water and transparent vessels, is an important tool to perform these studies in a laboratory scale at low cost and safe conditions [3]. Several groups of researchers have measured the size of the depth of the depression formed by the impingement of a gas jet on a liquid bath and they have obtained results in the form of empirical correlations of the dimensionless depth size as a function of the main experimental parameters [4]. Some of these correlations are obtained from measurements in pilot furnaces [3, 5] but most of them are obtained by physical modelling experiments [6, 7, 8, 9]. Additionally, the velocity profile in the bath induced by the momentum transfer from the gas jet is important in the mixing phenomena and in reactions kinetics governed by mass transport of species. In the literature there are works that describe recirculation in the liquid by mathematical models [10, 11, 12]. Regarding the cavity geometry, some researchers have reported results of the cavity depth [13, 14, 15]. Almost all the works reviewed performed rather simplified 2D simulations considering a vertical jet impinging on a free surface to compute both recirculation and cavity geometry. In this work a 3D mathematical model is developed using commercial CFD code and applying the VOF algorithm to simulate the impinging of a gas jet on a liquid bath. Numerical results of cavity geometry and recirculations in the liquid are compared against experimental measurements of velocity profiles in the region near the jet impingement obtained by using the PIV technique, and the cavity dimensions at different experimental conditions measured in images obtained using a high velocity camera.

NUMERICAL MODEL

Assumptions:
1. Incompressible and Newtonian fluid for both gas and liquid phases.
2. Velocity profile of the jet leaving the nozzle is considered to be uniform with a constant value.
3. Presence of a symmetry plane.

Governing equations

In the development of a mathematical model of a gas jet impinging on a free surface the Volume of Fluid (VOF) technique can be used. This technique is widely employed to describe the flow and the interface position separating two or more immiscible fluids [12, 15]. The free surface problem is approached as a single fluid flow problem where a conservation equation for a marker scalar variable, ξ, is solved. The value of the marker in every point in the domain defines the fluid present in that grid point and the properties of the fluid present in that location.

Marker conservation equation

The form of the marker conservation equation is given by eq. 1:

$$\frac{\partial \xi}{\partial t} + v \bullet \nabla \xi = 0 \qquad \text{(Eq. 1)}$$

where v is the fluid velocity vector and t is the time. In the VOF scheme, volume fractions of each one of the fluids present in the domain are determined in each cell. In this particular case, with two fluids in the domain, the value of the marker, ξ, is the volume fraction of the liquid phase in each node. Then, the fraction of gas is $1-\xi$ since the sum of all volume fractions in every cell must be one. With the ξ value defined at the cell node, the physical properties of the fluid at that node can be determined as a weighted average of the volume fractions of each phase. For example, the average density of the mix at the local position, ρ, can be computed as:

$$\rho = \rho_g (1 - \xi) + \rho_l \xi \qquad \text{(Eq. 2)}$$

where ρ_g is the gas density and ρ_l is the liquid density.

12

Momentum conservation equation

In its general vectorial form, the momentum conservation equation for a newtonian fluid can be written in the following form:

$$\frac{\partial}{\partial t}(\rho v) + \nabla(\rho v) = -\nabla(P) + \nabla[\mu \Delta(v)] + \rho g \qquad \text{(Eq. 3)}$$

where P is the pressure, g is the gravitational constant and μ is the average viscosity (calculated at each node in a similar way as the density through Eq. 2).

Turbulent flow

To describe turbulence in the system, the two-equation standard k-ε is employed [12,15].

General Conservation Equation

All conservation equations in the model developed in this work follow the form of the general conservation equation of the conserved quantity ϕ:

$$\frac{\partial(\rho\phi)}{\partial t} + \nabla(\rho v\phi) = \nabla(\Gamma_\phi \Delta\phi) + S_\phi \qquad \text{(Eq. 4)}$$

where Γ_ϕ is the diffusive coefficient and S_ϕ is the source term. The general conservation equation consists in the following terms from left to right in eq. 4: transitory term, convective transport of ϕ, diffusive transport of ϕ and the source term which is used to include all terms can not be inserted in the previous terms.

Boundary conditions

Figure 1 shows all boundaries of the computational domain, which is a half of a cylindrical vessel (due symmetry) containing a given amount of liquid reaching a determined level and the rest of the vessel is air. Boundaries are the cylindrical and bottom walls, a symmetry plane is also considered to save computer time and the opened surface at the top of the vessel.

1.- Walls

At the solid impermeable walls (cylindrical and bottom) the non slip condition is set for the velocities, and zero turbulence is established. Universal wall functions are used to describe the velocity profiles from the fully turbulent region to the laminar zone near the walls.

2.- Symmetry

Due to the geometrical features of the vessel and the location of the lance insufling the gas jet, it is possible to define a symmetry plane in the vessel. The symmetries are considered zero flux boundaries for both mass and momentum transfer.

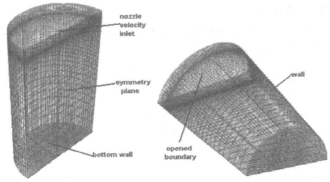

Figure 1.- Boundary conditions.

3.- Open boundary

This boundary is located at the top of the vessel and is opened to the atmosphere. Consequently, the pressure at this boundary is fixed to the atmospheric value. With this setting, flow in or out of the vessel is possible and the mass entering or leaving the domain depends on the mass balance.

4.- Gas jet inlet

At the nozzle tip the gas jet is assumed to have uniform velocity magnitude and the velocity components are set according to the lance angle, while turbulence at the nozzle tip is evaluated as follows:

a) Turbulent kinetic energy

$$k = 1.5 \left(V_m I \right)^2 \qquad\qquad \text{(Eq. 5)}$$

Where V_m is the magnitude of velocity at the nozzle and I is the turbulence intensity defined as:

$$I = 0.16 \left(Re \right)^{-1/8} \qquad\qquad \text{(Eq. 6)}$$

where Re is the Reynolds number.

b) Dissipation rate of turbulent kinetic energy

$$\varepsilon = 0.164 \frac{k^{3/2}}{l} \qquad \text{(Eq. 7)}$$

Where k is the turbulent kinetic energy and l is the turbulence length scale.

Numerical solution

The mathematical model is casted into a commercial fluid dynamics code and simulations are performed in a Work Station T3400, Core 2 Duo processor. A structured 3d non-uniform grid is developed to represent the computational domain. The numerical technique of VOF is used to track the free surface deformation due to the impingement of the jet, and the geo-reconstruct scheme is also applied. Interpolation of the pressure values at the cell faces using the momentum equation is performed with the PRESTO! (Pressure Staggering Option) scheme, and pressure and velocities are coupled using the PISO algorithm (Pressure-Implicit with Splitting of Operators). Transient computations are run until 10 seconds of real time is reached, despite the fact that results of the cavity geometry are stable at the first second of computation. All cases used a time step value of 10^{-4} seconds. Figure 1 shows an example of the computational mesh used (56 400 cells). The operational conditions for each case run to perform a process analysis are presented in Table I.

Table I.- Experimental conditions for each case.

Number of experiment	Nozzle diameter (mm)	Lance angle (degrees)	Gas flow rate (l/min)	Lance height (mm)
2				20
3	1.58	60	80	50
4				80
5				100
1			60	
3	1.58	60	80	50
8			100	
9			120	
3	1.58	60	80	50
6		75		
		90		
3	1.58	60	80	50
7	3.17			

RESULTS AND DISCUSSION

Cavity Geometry

Photographs are taken with all the experimental arrangements and then compared to mathematical results. Figure 2 shows photographs of typical cavities formed in the bath by varying the gas flow rate. Each cavity picture is compared against the cavity predicted through the mathematical model. We cannot include all experimental images here but the numerical results and the experimental measurements are presented in Table II.

In general terms, there is a good agreement between the measured and calculated dimensions. About cavity depth, the measured and calculated values are between 0.4 cm and 1.7 cm. It is clearly seen that as the gas flow rate increases, cavity depth and surface instability increase in both predicted and measured images of the cavity region. Figure 2 shows the comparison between the experimental and mathematical results by varying the gas flow rate in experiments 1

15

(60 l/min) and 9 (120 l/min). As distance from the nozzle decreases, cavity depth and surface instabilities increase. At different lance angles, 60° and 75° (and keeping the other variables as 1.58 mm of nozzle diameter, 50 mm of lance height and 80 l/min of gas flow rate), agreement is very good for cavity depth size although it is not that good with respect to the cavity width. The predicted cavity width has a larger diameter than the measured one. This is probably due to a overpredicted shear stress from the gas jet to the bath surface. The contribution of the shear stress is greater as the lance angle approaches to the horizontal (60°) but the jet pressure decreases, which results in a smaller cavity depth and in a higher momentum transferred from the jet to the bath by viscous shear. This promotes most likely more motion in the bath and a wider cavity. Overprediction of the model increases as the angle approaches to zero.

Table II. Experimental and mathematical values of diameter and depth of cavity, including the absolute values of % error of mathematical model.

No. of Exp.	Experimental Diam. (cm)	Model Diam. (cm)	Absolute % of Error	Experimental Depth (cm)	Model depth (cm)	Absolute % of Error
1	2.66	2.64	0.75	0.60	0.64	6.66
2	2.24	3.30	47.32	1.70	1.60	5.88
3	2.82	3.14	11.34	0.99	1.14	15.15
4	2.65	2.80	5.6	0.64	0.60	6.25
5	2.69	2.70	0.37	0.44	0.40	9.1
6	2.80	3.00	7.14	1.12	1.00	10.71
7	2.64	2.75	4.16	0.98	1.00	2.04
8	3.09	3.27	5.82	1.19	1.09	8.4
9	3.39	3.20	5.6	1.69	1.55	8.28

The results of the cavity geometry as a function of the nozzle diameter (two diameters, 1.58 and 3.17 mm, the rest of variables are lance height of 50 mm, gas flow rate of 80 l/min and lance angle of 60°) show an excellent agreement for the cavity depth and reasonable good for cavity width. As the nozzle diameter decreases (1.58 mm) the jet exits the nozzle with higher velocity and impinges on the bath with more strength causing a deeper cavity.

In order to further validate the model developed in this work, Figure 3a shows a comparison of the measurements (full symbols) and predictions (empty symbols) of the cavity depth together with a dimensionless semiempirical correlation (solid line) reported by Wakelin [1]. This semiempirical tendency has shown to be valid universally and it has been employed in previous work [16]. In the figure a dimensionless depression of the cavity $(H/z)*((H+z)/z)^2$ is plotted against the dimensionless number $J_z/(g\rho z^3)$, which involve the momentum of the jet (J_z) and the buoyant force.

Measurements and predictions of cavity depths are in excellent agreement with Wakelin's correlation when the value of the dimensionless number $J_z/(g\rho z^3)$ is below 0.006. This implies low momentums of the gas jets (J_z) producing small liquid surface instabilities, but above this value predicted and measured cavity depths show a poor agreement and a deviation with respect to Wakelin´s correlation.

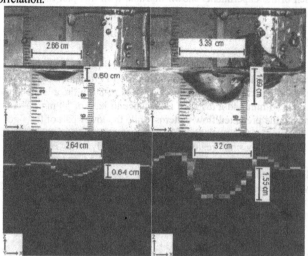

Figure 2. Experimental photographs and mathematical model results after varying gas flow rate: (a) Nozzle diameter 1.58 mm, gas flow rate 60 l/min, lance angle of 60° and lance height of 50 mm, (b) Nozzle diameter 1.58 mm, gas flow rate 120 l/min, lance angle 60° and lance height 50 mm.

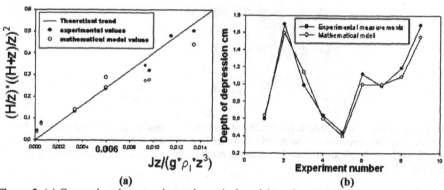

(a) (b)

Figure 3. (a) Comparison between the mathematical model results, presented in empty circles and the obtained values by using Wakelin correlation [1], in full circles and (b) depression depth results obtained with the mathematical model (empty circles) and compared with the experimental results (full circles).

17

Additionally in Figure 3b, there is a direct comparison between cavity depths predicted and measured for each of the experiments shown in Table I. In this plot the excellent agreement is evident between experiments and model calculations of cavity depressions.

Velocity Profiles

Figure 4 shows the velocity profiles predicted by the model and measured in the physical model. The experimental measurements are made by using the PIV technique in the region of the bath near the impinging point of the jet with a gas flow rate of 80 l/min, a lance angle of 60° with a nozzle diameter of 1.58 mm and a distance from lance to bath of 50 mm. The upper part of the figure corresponds to the predicted flow patterns while the lower part of the figure represents the experimentally measured velocity fields. The schematic drawing of the system at the right of Figure 4 shows the relatively small region of the domain analyzed by PIV.

Both measured and predicted flow patterns have the same order of magnitude of liquid velocities. They are typically found for a gas jet impinging on a bath. Liquid velocities are around 0.1 m/s where the jet impinges the bath and as distance increases axially and/or radially away from the impingement point, they tend to decrease.

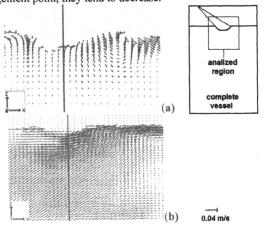

Figure 4. Obtained flow pattern. (a) Results of mathematical model and (b) measurements by PIV. Experimental conditions are: nozzle diameter 1.58 mm, gas flow rate 80 l/min, lance angle of 60° and lance height of 50 mm. The drawing on the right shows the analyzed region in the complete domain.

Figure 5 shows typical plots of the measured and predicted horizontal component of the velocity (x component) along the vertical distance from the bath surface to the bottom of the vessel. This corresponds to experiments 1 (Figure 5a) and 3 (Figure 5b. Agreement in almost all cases is reasonable good between model predictions and measurements. In all cases axial profiles of the x-velocity (horizontal component) are the same, i.e., at the free surface velocity is set according to the jet strength at the bath free surface and suddenly increases due to the shear action of the jet

18

until it reaches a maximum. Subsequently, velocities decrease asymptotically to zero near the bottom of the vessel. The maximum x-component of liquid velocity is around 2 to 4 cm from the original free surface. Flow patterns are important because they determine mixing of chemical species and thermal uniformity by convection in the real steelmaking processes.

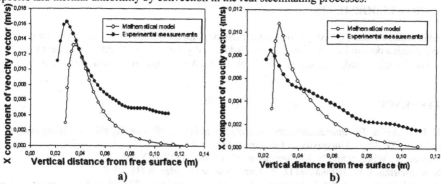

Figure 5. X component of the velocity vectors along a vertical line from the cavity bottom, varying nozzle diameter. Experimental measurements are in full symbols and mathematical model results are in white symbols. (a) Nozzle diameter 1.58 mm, gas flow rate 60 l/min, lance angle of 60° and lance height of 50 mm, (b) nozzle diameter 1.58 mm, gas flow rate 80 l/min, lance angle 60° and lance height 50 mm.

CONCLUSIONS

Computed cavity depths formed as a result of gas jet impinging on a liquid free surface satisfactorily agree with experimentally measured cavities for all experimental conditions considered in this work. Cavity depths computed with the mathematical model here developed are in full agreement with experimental measurements and Wakelin's empirical correlation for values equal or below 0.006 of the dimensionless number $J_z/(g\rho z^3)$. This corresponds to jets impinging on the free surface with low strength (low values of J_z). Above this value, agreement between predicted and experimental cavity depths is poorer and there is a deviation of cavities measured and computed with respect Wakelin's correlation.

Forces governing cavity geometries of a gas jet (air) impinging a liquid (water) free surface are inertial and buoyant forces under the narrow range of experimental conditions employed in this study.

Computed liquid flow patterns near the cavity agreed reasonably well in both trend and magnitude against experimentally determined flow patterns

Axial profiles of the horizontal component of liquid velocity from the cavity to the bottom of the vessel are successfully represented by the mathematical model since the predicted profiles show the same trend as the measured profiles, i.e. the horizontal component of the liquid velocity increases from the cavity until it reaches a maximum located around 2 to 4 cm from the original free surface and then it decreases asymptotically until zero at the bottom of the vessel.

ACKNOWLEDGEMENTS

Authors thank to the CONACyT to provide financial support for this work through Grant 60033, and through a doctoral scholarship to J.S-L. Additionally, Grant UNAM PAPIIT Project IN 109310-2is acknowledged. Authors also thank M. Sc. Adrian Amaro V. for his technical support.

REFERENCES

1 D. H. Wakelin: *The interaction between gas jets and the surfaces of liquids, including molten metals*. Ph. D. thesis, University of London, 1966.
2 M. Lee, V. Whitney and N. Molloy: *Scand. J. Metall.*, **30** (2001), 330.
3 S. K. Sharma, J. W. Hlinka and D. W. Kern: *Iron Steelmaker*, **4** (1977), 7.
4 A. Nordqist, N. Kumbhat, L. Jonsson and P. Jonssön: *Steel Res.*, **77** (2006), 82.
5 S. C. Koria and K. W. Lange: *Steel Res.*, **58** (1987), 421.
6 F. Qian, R. Muthasaran and B. Farouk: *Metall. Mater. Trans. B*, **27B** (1996), 911.
7 Subagyo, G. A. Brooks, K. S. Coley. and G. A. Irons: *ISIJ Int.*, **43** (2003), 983.
8 F. Memoli, C. Mapelli, P. Ravanelli and M. Corbella: *ISIJ Int.*, **44** (2004), 1342.
9 A. R. N. Meidani, M. Isac, A. Richardson, A. Cameron and R. I. L. Guthrie: *ISIJ Int.*, 44 (2004) 1639.
10 K. D. Peaslee and D. G. C. Robertson: EPD Congress, TMS, Warrendale, PA, 1994, 1129.
11 L. Gu and G. Irons: Electric Furnace Conf. Proc., Iron & Steel Soc., Pittsburgh, PA, 1999, 269.
12 M. Ersson, A. Tilliander, L. Jonsson and P. Jönsson: *ISIJ Int.*, **48** (2008), 377.
13 M. P. Schwarz, Fluid Flow Phenomena in Metals Processing, ed. by N. El-Kaddah, D. G. C. Robertson, S. T. Johansen and V. R. Voller, TMS, Warrendale, PA, 1999, 171.
14 D. Nakazono, K. Abe, M. Nishida and K. Kurita: *ISIJ Int.*, **44** (2004), 91.
15 A. V. Nguyen and G. M. Evans: *Appl. Math. Model.*, **30** (2006), 1472.
16 J. Solórzano-López, R. Zenit and M. A. Ramírez-Argáez: *Rev. Metal. Madrid* , (2010), Accepted for publication.

Mater. Res. Soc. Symp. Proc. Vol. 1276 © 2010 Materials Research Society

Manufacturing of Hybrid Composites and Novel Methods to Synthesize Carbon Nanoparticles

Andres E. Fals, Julio Quintero, F. C. Robles Hernández[1]
University of Houston, Department of Mechanical Engineering Technology, College of Technology, 304 Technology Building, Houston, TX 77204-4020.

ABSTRACT

In following are presented the characterization results of nanostructured hybrid composites using alumina matrix reinforced with nanostructured particles of Ni, Ti and soot. The soot used in this work is the byproduct from the synthesis of Carbon Nanotubes (CNT) or fullerene and contains traces (<1wt%) of either CNT or fullerene. Ni and Ti are used in this work for their inherent catalytic ability for heterogeneous nucleation of carbon nanostructures (nanotubes, fullerenes). The hybrid composites are produced by a combination of methods including mechanical milling, sonication, and Spark Plasma Sintering (SPS). Mechanical milling is conducted in high energy mills, the milled and as manufactured powders are sonicated to assure their dispersion, homogeneity and promote percolation of the components during sintering. Mechanical milling and SPS have positive effects to promote the synthesis of different carbon nanoparticles. For instance, it is observed that mechanical milling of fullerene soot sponsors the synthesis of nanostructured diamond particles and using SPS can be synthesized diamond too and fullerene. Although, it is important to notice that SPS conditions are critical to the amount and type of synthesized particles. The use of CNT soot sponsors porosity, hence lower density resulting in an ideal material for membrane and porous media applications. The results of characterization (X-Ray diffraction, electron microscopy (scanning and transmission)) and the mechanical properties (Vickers microhardness) are discussed accordingly.

INTRODUCTION

The synthesis of fullerene (C_{20}, C_{60}, C_{70}, etc.) and CNT is possible by various methods resulting in interesting structures from buckyballs, buckytubes, giant fullerene structures, to concentric structures (e.g. onions), etc. [1-5]. The most common structures (buckyballs) are C_{60} and C_{70} that consists of atomic arrangements of carbon with pentagons, hexagons and in some cases heptagons [2,6]. It has been reported that fullerenes (C_{60}, C_{70}, etc.) and in general carbon nanostructures, such as carbon nanotubes (CNT), have outstanding mechanical properties with Young Moduli between 400 GPa and 4150 GPa [1,7] and compressive strengths between 100 and 150 GPa [1,8]. Some identified uses of fullerenes are as a unique reinforcement for structural materials and nanostruders [9-14].

A major goal in the development of ceramic matrix composites is to increase their toughness. The most successful efforts on this area for alumina matrix composites involve the use of carbon nanotubes (single-walled, multi-walled, and ropes) [15-18]. Apparently, the use of single walled carbon nanoropes seems to be the most successful route to improve fracture toughness in alumina matrix composites [18]. Mechanical milling is widely used to manufacture composites due to its ability to improve homogeneity, integrity and mechanical properties as a

---

1 Contact Author: fcrobles@uh.edu, tel. (713)-743-8231, fax. (505)213-7106.

result of microstructure refinement resulting in nanostructured composites [19,20]. In order to preserve the nanostructured nature of the mechanically alloyed powders, low temperatures, short times and high pressures are ideal for the corresponding sintering method. These conditions are combined in the Spark Plasma Sintering (SPS) technique making it one of the most promising methods to produce highly or fully dense nanocrystalline composites [10,11,12]. Some of the challenges to produce nanostructured ceramic matrix composites reinforced with carbon nanoparticles are: segregation, poor percolation, and cost.

The above has motivated the present research work in exploring the potential of carbon soots (fullerene, nanotube, graphene, etc.) instead of highly pure fullerene and CNT manufactured by means of mechanical alloying and SPS. The intention of this work is to demonstrate the potential of using carbon soot (fullerene and CNT) as an effective reinforcement and their inherit potential to transform into other carbon nanostructures during cold and hot processing. It means in-situ synthesis of carbon nanostructures to increase the mechanical properties and density of alumina matrix composites. In order to achieve this in the present work are combined mechanical alloying with SPS to produce alumina matrix nanostructured composites reinforced carbon nanostructures. The characterization is conducted by means of Scanning Electron Microscopy (SEM), X-Ray Diffraction (XRD) and Transmission Electron Microscopy (TEM), hardness, specific gravity and porosity results are also reported.

EXPERIMENT

Commercial alumina (Al_2O_3) powders (98 wt% purity, mesh 140, all particles < 106 μm) have been obtained from Sigma Aldrich. The carbon soot has been obtained from the SERES Company with traces of fullerene (C_{60} and C_{70}). The CNT soot was donated by the CIMAV-Chihuahua and has less than 2wt% CNT. Nickel powder had been obtained from Alfa Aesar with a nominal purity of 99.9 wt% and a particle size of 3 - 7 μm. Titanium powder has been purchased from Atlantic Equipment Engineers (99.7 wt % Ti) and a particle distribution below 74 microns (mesh 200). Electron micrographs of the powders in as obtained conditions are presented in Figure 1.

Mechanical milling has been conducted on a high energy system (SPEX) for 0, 0.5, 2, 5, 10, 20 and 50 h for the fullerene soot; the same conditions have been used for the Ni and Ti powder mix. In both cases have been used steel hardened mills. The alumina used in the present work has been milled for 20 h in the above mentioned system. No milling is induced into the CNT soot to prevent mechanical damage to the CNT. Following the mechanical milling all the powders are sonicated to break any physical contact among powders and to separate them into smaller particles. **Figure 2** shows a series of micrographs of the powders in the as milled and/or sonicated conditions.

The milled powders have been sintered under the following SPS conditions: temperature 600 °C, 1300 °C and 1500 °C and for 0 to 600 s and a heating rate of 140 °C/s. The average dimensions of the sintered samples are 12 mm in diameter and 4 mm in thickness. The sintered products have been used for microhardness and density measurements. The micro-hardness test has been conducted on a Future-Tech FX-7 Microhardnes Tester FM apparatus with an applied load of 1 kN for 15 s. The reported values are the average of 9 measurements made in the alumina matrix. The specific gravity measurements have been conducted using the conventional Archimedes method with de-ionized water at room temperature.

X-ray diffraction (XRD) has been conducted on a SIEMENS Diffractometer D5000 equipped with a Cu tube and a characteristic $K_\alpha = 0.15406$ nm operated a 40kV and 30 A. The SEM observations have been carried out on a FEI XL-30FEG using secondary electrons. The TEM/HRTEM has been conducted on a JEOL JEM200FXII. The milled and as synthesized powders are sonicated on a Misonix apparatus operated at 10-11 Watts for 30 min.

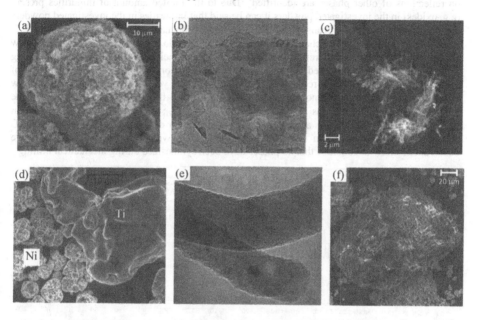

Figure 1. (a,c,d,f) scanning and (b,e) transmission electron micrographs for the (a,b) fullerene soot, (c,e) CNT soot, (d) Ni and Ti and (e) alumina powders in the as manufactured conditions.

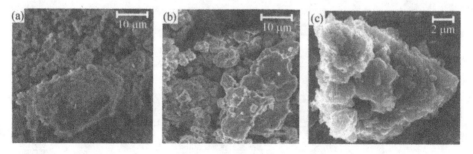

Figure 2. Scanning electron micrographs for milled powders of (a) Al_2O_3 for 10 h, (b) Ni-Ti for 50 h and (c) fullerene soot milled for 50 h.

RESULTS

Figure 3 shows the XRD result of the original powders used to manufacture the hybrid composites. The XRD results show that the original powders have a relatively high purity since no reflections of other phases are identified. Due to the limited amount of impurities present (e.g. oxides) in the investigated powders it can be said that impurities in the investigated powders are untraceable by XRD means. The XRD results of mechanical milling of the Ni and Ti powders mix indicate that mechanical milling of up to 50 h do not induce the synthesis of new phases; nonetheless, the XRD reflections become wider and their intensity decrease with milling time. Similar effect is observed with the alumina as a function of milling time. After 10 h of milling is observed a shift on reflections $(202)_{Ti}$ and $(111)_{Ni}$; yet, there are no presence of new reflections corresponding to another phase. Milling times 5 h or more cause a switch in the relative intensity of reflections $(111)_{Ni}$ reflection and $(202)_{Ti}$; it means the $(20\bar{2})_{Ti}$ reflection becomes more intense. This is attributed to the larger number of slip systems in Ni that is FCC and has 12 slip systems. On the other hand, Ti is HCP with 2 slip systems. The above observations allow to conclude that no phases have been synthesized during mechanical milling.

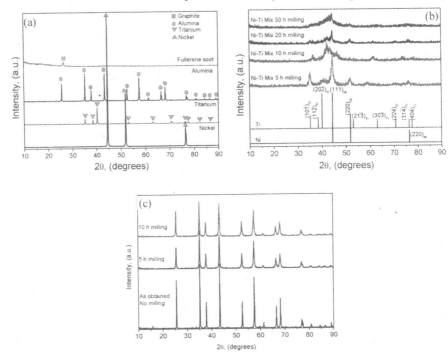

Figure 3. XRD spectrum of the powders used in the composited in the as received conditions.

The XRD results of the mechanical milled fullerene soot powders are presented in Figure 4. The main reflection in the fullerene soot is the $(002)_{Gr}$ peak. The presence of the fullerene peaks (most likely C_{60}) is attributed to the remaining C_{60} present in the soot after the C_{60} is removed for purification and is within 1wt% or less as indicated by the manufacturer. For milling times longer than 2 hr the presence of a new family of reflections between the 2θ degrees from 40 to 47 degrees is observed and these reflections become clearer as milling time increases. These reflections correspond to synthetic diamond with a hexagonal structure type P6₃/mmc and as reported in reference [21,22]. In the XRD spectrums of the fullerene soot milled for 50 h are identified Fe_3C reflections. The presence of Fe is attributed to the milling media (steel mill and balls) that presumably formed the Fe_3C compound. Further, by comparing Figure 1a and Figure 2c can be observed that milling has noticeable effects on the morphology of fullerene soot.

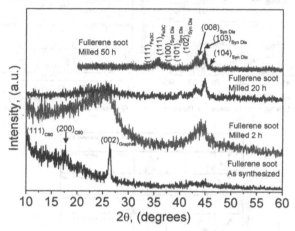

Figure 4. XRD of the as synthesized and mechanical milled fullerene soot powders.

Figure 5 shows the XRD results of the sintered aluminum matrix composite samples using fullerene soot or CNT soot as reinforcements. Figure 5a and b depict the effects of SPS processing on the synthesis of different type of carbon nanostructures. The presence of diamond, graphite and fullerene correspond to carbon nanostructures synthesized during the SPS process. The SPS sintering conditions have a direct influence on the type of synthesized carbon structures; for instance, relatively low temperatures (600°C) sponsor the synthesis of fullerene (likely C_{60}). At 1300 °C and 1 min endorses the synthesis of diamond, at the same temperature but for 10 min diamond presumably transform into graphite the rest o the carbon in the quasi-amorphous soot transform into graphite. The synthesized diamond or fullerene also depends on the amount of fullerere soot added to the composites. For instance, additions of 2wt% of fullerene soot sintered at 1500 °C for 1 min seems to be the most effective combination of parameters to synthesize diamond. On the contrary, using 1 wt% fullerene soot and sintering at 1300 °C for 10 min is a more effective way to synthesize graphite.

The results of the composites manufactured with CNT soot as reinforcement are presented in Figure 5c. The effect of the CNT soot during SPS sintering is different than that of fullerene soot. The composites manufactured with CNT soot reinforcements are produced with the CNT soot in the as manufactured conditions (no spex milling); instead, in order to disperse the CNT soot particles, the powders are sonicated. Thirty minutes of sonication on the CNT soot powders seems to be the ideal time to obtain a good dispersion of the powders reducing the amount of synthesized graphite during SPS. Furthermore, adding CNT soot and up to 5wt% of fullerene soot seem to be the ideal conditions to hinder the synthesis of diamond.

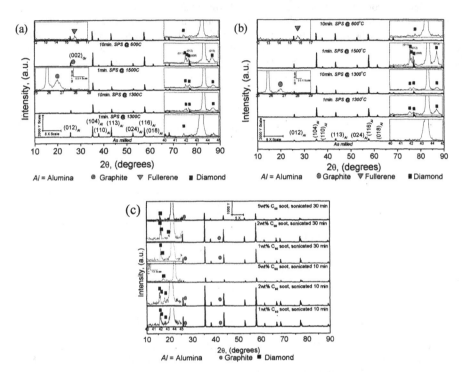

Figure 5. XRD results of the alumina matrix composites after SPS sintering using 2 wt% of milled Ni-Ti powders and (a,b) fullerene (C_{60}) soot or (c) CNT soot sintered under different SPS and reinforcement additions of a) 1wt% C_{60} soot, b) 2wt% C_{60} soot and c) 2wt% CNT soot.

Figure 6a-c presents the results of hardness, porosity and specific gravity for the composites treated under different SPS conditions. **Figure 6a** depicts that the composite sintered at 1500°C for 1 min shows that is the hardest composites produced in this work. The hardness reached by this composite is 14.6 GPa that is between 4.3 and 2.1 times higher that values reported for alumina membranes [23]. The hardness of the composites presented in this work is between 79 % and 97 % of the hardness previously reported for highly pure/highly dense alumina composites [24]. The reason for the slightly lower hardness of in the composites presented in this work is mainly attributed to the difference in grain size as well as the presence of carbon particles. On the contrary, when using CNT soot the hardness is remarkably lower in all SPS cases SPS conditions and this hardness is comparable to that of porous membranes (5.61 GPa) or and lower 2.45 GPa [23].

Figure 6b presents the results of porosity where can be observed that a temperature as low as $0.75T_m$ of that of melting in absolute scale (homologous temperature) is enough to reach full densification of the composites for samples with 1-5wt% fullerene soot. It is interesting that sonication time makes a significant different to disperse the samples allowing the breakage of the agglomerates in the as milled products and this results in better contact among the constituents of the composite increasing percolation. The use of CNT soot has positive effects in porosity having positive effects on lowering the density of the composite. This allows concluding that temperature is the main factor affecting density followed by sonication.

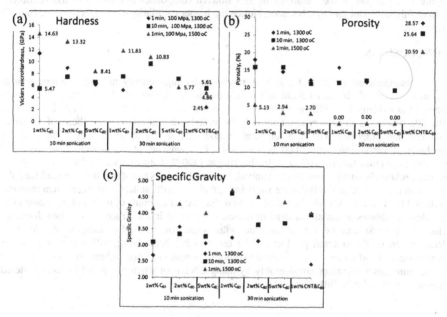

Figure 6. Results of mechanical (a) and physical (b,c) characteristics of the sintered composites; a) hardness, b) porosity and c) specific gravity.

Figure 7 depicts examples of SPS sintered samples for composites reinforced with fullerene soot or CNT soot. Figure 6 shows the results of the specific gravity and is evident that temperature is the main factor influencing the level of densification of the composites. However, the use of small amounts of fullerene soot (1wt%) and sonication (30 min) has positive effects on density. On the other hand using CNT can be achieved porous materials with lower density that can be idea for membranes and porous media applications. On the contrary, higher densification of the material reduce the capillarity of the composite reducing its wetability that is mainly attributed to the higher density of the fullerene soot reinforced composite.

27

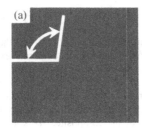

Figure 7. Examples of the wetability of SPS sintered composites of alumina matrix reinforced with a) fullerene soot and b) CNT soot. Notice the higher capillarity of the CNT soot reinforced composite.

CONCLUSIONS

Nanostructured diamond can be synthesized by high energy milling fullerene soot at "room temperature". In the mechanically milled product the transformation of the carbon soot to diamond is evident; for SPEX milling times of 50 h or more the presence of Fe_3C is observed. The SPS process is an ideal method not only for sintering but also for the synthesis of diamond, fullerene and graphite. The SPS temperature and time seems to be the key factors for the synthesis of carbon nanostructured compounds. For instance, at a temperature of 600°C is possible to produce fullerene (C_{60}). SPS sintering at 1300 °C for 1 min are the ideal conditions to produce relatively small amounts of diamond, longer SPS times transform the diamond (and the other carbon present in the fullerene soot) into graphite. SPS sintering at higher temperatures (1500°C) for 1 min are the ideal conditions to produce larger amounts of diamond that seem to be the ideal conditions to transform most of the carbon present in the fullerene soot into diamond. The SPS process can be further assisted with sonication to further increase the level of densification of the sintered products. The use of CNT soot during SPS sintering at lower temperatures (1300°C) for 1 min are ideal to produce porous media (membranes). The capillarity of the composites increases considerably by using CNT soot when compared to these sintered products produced with fullerene soot.

ACKNOWLEDGMENTS
The authors would like to expresses their gratitude to Dr. D. Butt for facilitating the sintering of the samples at Center for the Advanced of Energy Studies (CAES) at Idaho Falls. Special thanks to the University of Houston and the government of Texas for the Start Up and HEAFS funding. Special thanks to Dr. G. Majkic at the Mechanical Engineering Department (University of Houston) for his assistance to facilitate the microhardness equipment.

REFERENCES
1. P. J. F. Harris, Carbon nanotubes and related structures, Cambridge University Press, Cambridge 1999.
2. H. W. Kroto, J. R. Heath, S. C. O'Brien, R. F. Curl, R. E. Smalley, C60: Buckminsterfullerene, Nature, **318,** 162-163 (1985).
3. S. Iijima, Helical microtubules of graphitic carbon, Nature, **354,** 56-58 (1991).

4. Q. Ru, M. Kamoto, Y. Kondo, K. Takayanagi, Attraction and orientation phenomena of bucky onions formed in a transmission electron microscope, Chem Phys Lett **259**, 425-431 (1996).
5. D. Ugarte, Curling and closure of graphitic networks under electron-beam irradiation, Nature, 359 (1992) 707-709.
6. H. Terrones, M. Terrones, The transformation of polyhedral particles into graphitic onions, J Phys Chem Solids, **58** 1789-1796 (1997).
7. M. M. Treacy, T. W. Ebbesen, J. M. Gibson, Exceptionally high Young's modulus observed for individual carbon nanotubes, Nature, **381**, 678 (1996).
8. O. Lourie, D. M. Cox, H. D. Wagner, Buckling and Collapse of Embedded Carbon Nanotubes, Phys Rev Lett., **81**, 1638-1641 (1998).
9. M. Umemoto, K. Masuyama, K. Raviprasad, Mechanical Alloying of Fullerene with Metal, Mats. Sci. Forum **47**, 235-238 (1997).
10. F. C. Robles Hernandez, Production and Characterization of Composites Metal-C (where Metal= Fe or Al and C= Graphite or Fullerene) Obtained from Mechanically Alloyed Powders, MSc Thesis, Instituto Politécnico Nacionál, Departamento de Ingeniería Metalúrgica, Mexico, 1999.
11. F. C. Robles Hernández, H. A. Calderón, "Nanostructured Metal Composites Reinforced with Fullerenes", JOM, 62, 2, 2010, 63-68.
12. F. C. Robles Hernández, H. A. Calderón, "Synthesis of fullerene on Fe-C composites by Spark Plasma Sintering and its thermomechanical transformation to diamond", MRS Proceedings, 1243, 2010.
13. L. Sun, F. Banhart, A. V. Krasheninnikov, J. A. Rodrguez-Manzo, M. Terrones, P. M. Ajayan, Carbon Nanotubes as High-Pressure Cylinders and Nanoextruders, Science, **312** 1199-1202 (2006).
14. V. Garibay-Febles, H. A. Calderon, F. C. Robles-Hernández, M. Umemoto, K. Masuyama, J. G. Cabañas-Moreno, Production and Characterization of (Al, Fe)-C (Graphite or Fullerene) Composites Prepared by Mechanical Alloying, Mats & Manufac Processes, **15**, 547-567 (2000).
15. G-D. Zhan, J. Kuntz, J. Wan, J. Garay, A. K. Mukherjee, "A Novel Processing Route to Develop a Dense Nanocrystalline Alumina Matrix (<100 nm) Nanocomposite Material, J. Am. Ceram. Soc., **86**, 200–202 (2002).
16. G-D. Zhan, J. Kuntz, J. Wan, J. Garay, A. K. Mukherjee, Alumina-based nanocomposites consolidated by spark plasma sintering, Scripta Materialia 47, 737–741 (2002).
17. E. Zapata-Solvas, D. Gomez-Garcia, R. Poyato, Z. Lee, M. Castillo-Rodriguez, A. Dominguez-Rodriguez, V. Radmilovic, N. P. Padture, Microstructural Effects on the Creep Deformation of Alumina/Single-Wall Carbon Nanotubes Composites, J. Am. Ceram. Soc., 93, 7, 2042−2047 (2010)
18. G-D. Zhan, J. D. Kuntz, J. E. Garay, A. K. Mukherjee, Electrical properties of nanoceramics reinforced with ropes of single-walled carbon nanotubes, **83**, 6, 1228-1230 (2003).
19. P.S. Gilman and J.S. Benjamin, Ann. Rev. Mater. Sci. **13**, 279 (1983).
20. C. Suryanarayana, Prog. Mater Sci. **46**, 1, (2001).
21. P. D. Ownby, X. Yang, J. Liu, Calculated X-ray Diffraction Data for Diamond Polytypes, J Am. Ceram. SOC., **75**, 7 (1992) 1876-83.

22. I. Santana-García, F.C. Hernandez-Robles, V. Garibay-Febles, H. A. Calderon Metal (Fe, Al)-Fullerene Nanocomposites: Synthesis and Characterization, Microscopy and Microanalysis, 16, S2 (2010) 1688-1689.
23. Z. Xia, L. Riester, B. W. Sheldon, W. A. Curtin, J. Liang, A. Yin, J. M. Xu, Mechanical properties of highly ordered nanoporous anodic alumina membranes, Rev. Adv. Mat. Sci. 6, 2004, 131-139.
24. A. Krell, P. Blank, Grain size dependence of hardness in dense sub micrometric alumina, J. Am. Ceram. Soc., 78, 4, (1195) 1118-1120.

Mater. Res. Soc. Symp. Proc. Vol. 1276 © 2010 Materials Research Society

Effect of the deposition rate on thin films of CuZnAl obtained by thermal evaporation

L.López-Pavón[1], E. López-Cuellar[1,2], A. Torres-Castro[1,2], C. Ballesteros[3], C. José de Araújo[4].

[1]FIME-UANL, Ave. Universidad S/N. Cd. Universitaria, San Nicolás de los Garza, Nuevo León, México. C.P. 66450
[2]CIIDIT, Km. 10 de la Nueva Autopista al Aeropuerto Internacional de Monterrey, Apodaca, Nuevo León, C.P. 66600
[3]Departamento de Física, Universidad Carlos III de Madrid, Avda. Universidad 30, 28911 Leganés, Madrid. Spain.
[4]Department of Mechanical Engineering, Universidade Federal de Campina Grande, Av. Aprígio Veloso, 882, Bodocongó, Campina Grande - PB, Brazil.

ABSTRACT

Thermal evaporation is used to deposit thin films of CuZnAl on silicon substrates. For this purpose, a CuZnAl shape memory alloy is used as evaporation source. The chemical composition and the phases present in the films are evaluated at two different deposition rates: 7 and 0.2 Å/s. The thin films are heat treated to promote the diffusion of the elements and characterized by X-ray Diffraction, Energy Dispersive X-ray Spectroscopy and Scanning Transmission Electron Microscopy (STEM). It is shown that the chemical composition of the thin films is significantly different to that of the CuZnAl alloy used as evaporation source. Moreover, the films produced at 7 Å/s show a significant loss of Zn, contrary to the results obtained using a deposition rate of 0.2 Å/s. It is also observed that the composition varies across the thickness of the film, suggesting that the various alloying elements are evaporated at different rates during the deposition process. Finally the predominant phases present in the films belong to the Al_xCu_y family.

INTRODUCTION

In recent years, the study of nanoparticles or the deposition of thin films of shape memory alloys has become a field of great interest for scientists due to their potential to become a primary actuating mechanism for micro-actuators and biomedical applications [1]. An alloy is considered as a Shape Memory Alloy (SMA) when it can 'remember' its shape, that is, after a SMA sample has been deformed from its original shape, it regains its original geometry by itself during heating (shape memory effect) or, simply during unloading at a higher ambient temperature (superelasticity) [2]. These properties are associated to a reversible, temperature dependent solid state martensitic phase transformation from a low-symmetry (martensite) to a highly symmetric crystallographic structure (austenite) [3]. SMA has a lot of interesting applications; there are more than 10,000 patents on this matter [4-5]: connectors, thermal or electrics activators, superelastic products, dampers and intelligent materials. Shape memory alloy thin films have been widely used in micro-electro-mechanical systems (micro-actuators) such as micropumps or microvalves [6-13].

The physical vapour deposition (PVD) techniques can be divided into two groups: evaporation and sputtering. In the present study the interest is in thermal evaporation, where the growing species are removed from the source by the application of thermal energy. Thermal evaporation has been used to fabricate thin films from different materials. Basically, the simplest process consists in heating the bulk target with an electric current in a high vacuum chamber. The target material reaches the vapour phase and is condensed on the substrate, which is maintained at a lower temperature. The material can be evaporated from liquid phase or directly from solid phase depending on pressure and temperature conditions.

In this work, thin films are fabricated by thermal evaporation using a CuZnAl shape memory alloy as the evaporation source. In order to study the effect of the deposition rate on the chemical composition and the phases present on the films, two different deposition rates were studied: 7 and 0.2 Å/s. The compositional variations and phases present on the thin films are analyzed and discussed.

EXPERIMENTAL

The targets or evaporation sources are made of a CuZnAl shape memory alloy. This material exhibits a martensitic transformation start temperature (Ms) of 246K. These targets are evaporated in an Angstrom Amod Deposition System under a high vacuum ($<10^{-6}$ Torr). The source material is heated by an electrical current proportional to the deposition rate (Å/s). In order to explore the effect of the deposition rate two different rate values were used: 7 and 0.2 Å/s. These deposition rates were selected to avoid problems of thickness and chemical composition control at very high or very slow deposition rates.

Silicon wafers with a [100] orientation are used as deposition substrates. The produced samples are characterized before and after a 1 hour-heat treatment carried out at 720 °C under an Ar atmosphere followed by water quenching. Scanning electron microscopy (SEM-FEI Nova nanosem 200), high resolution transmission electron microscopy (HRTEM, Tecnai 20F FEG) and X ray diffraction (XRD, Philips X'Pert MPD diffractometer) are used for structural and microstructural characterization.

RESULTS AND DISCUSSION

The chemical composition of the targets was evaluated by a semiquantitative EDXS analysis in the SEM. The results presented in Fig. 1 show the presence of 74.83 % wt Cu, 7.01% wt Al and 18.16%wt Zn. This composition corresponds to an alloy with an Ms of 246 K.
Thin film deposited at 7 Å/s

Figure 2 shows bright field TEM images of the deposited film before (Fig. 2A) and after (Fig. 2B) heat treatment.. Prior to heat treatment, a discontinuous layer (grey contrast) of grains (size < 500 nm) is observed together with a number of bigger grains of ~1μm in size (Fig. 2A). After the heat treatment, a continuous gray matrix with a higher density of dark grains of about 1μm in size is observed (Fig. 2B).

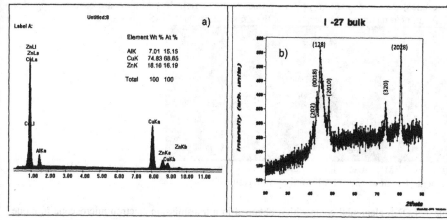

Figure 1.- A) Compositional semi-quantitative analysis performed in a SEM, B) XRD diffraction patterns obtained with the martensitic phase.

When the composition of the thin film is analyzed by EDXS, no clear Zn signal can be detected, as shown in Table 1. The reason for the absence of Zn in the EDXS spectra is that, at the deposition rate of 7 Å/s, the electrical current and the vacuum conditions cause Zn evaporation prior to evaporation of any other element (Zn has the lowest boiling temperature in the SMA composition). As a result, most of the zinc is deposited on the shield that protects the substrate before reaching the 7 Å/s deposition rate condition. This phenomenon is found in other alloys containing Zn and is called "dezincification" [14-17]. A small increase in Al was observed in the composition of the film heat treated; this is probably due to the diffusion of the aluminum up to the surface.

The analysis of the crystalline phases present the heat treated films was carried out by electron diffraction in the HRTEM (Fig. 3a). XRD was used to characterize the phases present in samples before and after the heat treatment (Fig. 3b). The spots in the electron pattern of Fig. 3a correspond to the Si substrate while the rings fit with the $AlCu_4$ phase. These results are confirmed by the XRD results as shown in Fig. 3b. The peaks correspond also to the Si-substrate and $AlCu_4$. These results indicate that thin films produced using a deposition rate of 7 Å/s are composed mainly by the $AlCu_4$ phase with no evidence of the presence of Zn.

Thin film deposited at 0.2 Å/s

With the objective of avoiding the dezincification effect observed at the high deposition rate, the lowest controlled rate recommended for the Angstrom Deposition System of 0.2 Å/s was used. Figs. 4a, 4b and 4c illustrate the morphology (SEM images) of three different films with thickness of 250, 300 and 400 nm, prior to the heat treatment. The EDXS elemental chemical composition of the films is given inside the micrographs. The microstructure of the 250 nm thick film (Fig. 4) consists of Zn-rich very fine spherical (white dots) and large needle-like particles (white lines) As the deposition process proceeds, these rich Zn particles are covered (Fig. 4b) with Al and Cu resulting in a decrease of the Zn signal in the EDXS spectra.

Figure 2.- TEM images of film deposited at 7 Å/s: A) before heat treatment, B) after heat treatment (1 hour at 720 °C followed by water quenching.

Table 1. EDXS elemental composition obtained in a transmission electron microscope in mode STEM.

%	Without HT		HT	
	Wt	At	Wt	At
Cu	89.16	77.74	86.17	72.57
Zn	--	--	--	--
Al	10.84	22.26	13.83	27.43

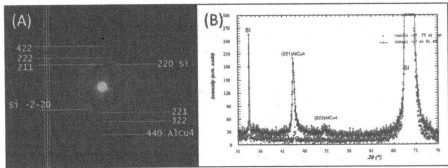

Figure 3 A) Electron diffraction and B) X-Ray diffraction patterns showing the presence of Si and AlCu₄ phase.

This result is consistent with the composition measured for the 400 nm thick film (Fig. 4c). As the thickness increases, the Zn content decreases due to its lowest boiling temperature. In contrast, the Cu content in the film increases. Moreover, the appearance of some "canyons" can be observed in the 400 nm thick film. In this case, the Zn content of lower zones are rich in Zn while higher zones are poor in Zn. This deposition behavior has the Wolmer-Weber form and is commonly called "island deposition" [18].

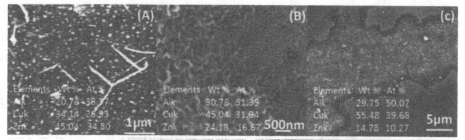

Figure 4.- Image of three different layers, where the dezincification its present. a) Thin film deposited of 250 nm thickness, b) 300 nm of thickness and c) 400 nm of thickness.

Figure 5 shows the variation of microstructure and EDXS elemental chemical composition through the cross-section of a thin film after the heat treatment. The image on the left of the figure shows the Si substrate, the thin film and the epoxy resin (white arrows) located on both sides of the film. As can be seen, the elemental distribution of Cu, Al and Zn is not uniform. . After the heat treatment, the Zn appears segregated to the thin film-substrate area on the right and left sides of the image. In contrast, Cu and Al appear uniformly distributed on the central part of the image. This elemental distribution suggests the formation of different phases on the thin film after the heat treatment.

When the sample is analyzed by TEM, a polycrystal with several types of grains is observed (Fig. 6). This observation also suggests the presence of different phases. Fig. 6c shows a high resolution image of twins observed in the microstructure of the film. The Fast Fourier Transform (FFT) simulated diffraction pattern of the zone enclosed by the white square in Fig. 6c is shown in Fig. 6d. The interatomic distances estimated at the resolution achieved, correspond to the Al_4Cu_9 phase. To analyze the phases present in a larger area, Electron Diffraction and X-Ray Diffraction (XRD) were carried out on the sample (Figs. 7a and 7b, respectively.. Two phases can be indexed in the electron diffraction pattern: the ring pattern corresponds to the $CuAl_2$ phase and the spot pattern to the Al_4Cu_9 phase, which is in agreement with the FFT results. The XRD results show peaks of several phases: $AlCu_3$, Al_4Cu_9, $CuAl_2$, Cu and Al.

In summary, results show that the chemical composition of the thin film produced by evaporation at a deposition rate of 0.2 Å/s contains all three elements (Cu, Al and Zn) present in the the SMA alloy used as evaporation source. In contrast, increasing the evaporation rate to 7 Å/s produces thin films with no Zn in their chemical composition. Thus, when a shape memory alloy is used as evaporation source, the individual elements appear to be evaporated at a different rate depending on their boiling temperature. This effect results in variations of chemical composition in the films both in plane and across the film thickness.

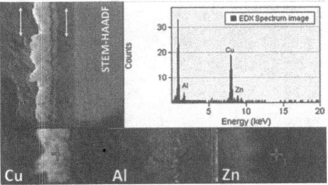

Figure 5. Cross section of a heat treated thin film and chemical composition, where is possible the detection of the carbon, aluminium and zinc are detected by EDX-mapping in STEM mode.

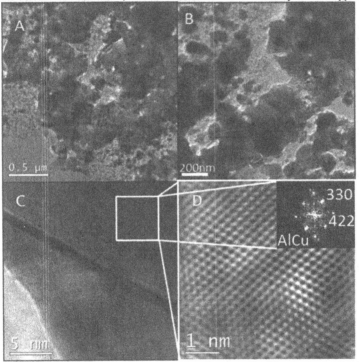

Figure 6, A) Micrograph of several types of grains, B) different sizes of grain where is possible to see the twins in the grains. C) High resolution image where high order crystalline areas are present (D) Fourier transform filtered image taken from the indicated selected zone. The insets in the figure (d) show the indexed Fourier transform diffraction pattern simulation.

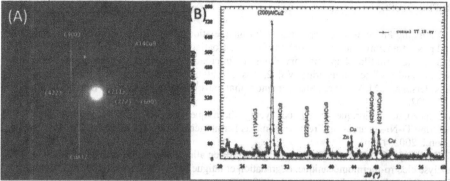

Figure 7.-A) Electron diffraction of two different phases, the rings corresponds to CuAl2 and the spots to the phase Al4Cu9. B) X-rays diffraction of the different phases present in an alloy CuZnAl.

In summary, results show that the chemical composition of the thin film produced by evaporation at a deposition rate of 0.2 Å/s contains all three elements (Cu, Al and Zn) present in the the SMA alloy used as evaporation source. In contrast, increasing the evaporation rate to 7 Å/s produces thin films with no Zn in their chemical composition. Thus, when a shape memory alloy is used as evaporation source, the individual elements appear to be evaporated at a different rate depending on their boiling temperature. This effect results in variations of chemical composition in the films both in plane and across the film thickness.

CONCLUSIONS

Thin films of different compositions were obtained by a thermal evaporation process using deposition rates of 0.2-7 Å/s. Results suggest that when an alloy source is thermally evaporated, each element will be evaporated at a different rate depending on their boiling temperature, elements with lower boiling temperatures evaporate first. In the present case of a CuZnAl SMA alloy, evaporation occurs in the following order: Zn, Al and Cu. This effect is more pronounced when the deposition rate is increased. Due to the difference in boiling temperatures and equilibrium vapour pressures of the elements in the SMA alloy, it is difficult to control the chemical composition of the films produced. Nevertheless, the results of the present work show that Zn can be retained in thin films deposited at a low rate. This is very important for future investigations because it opens the possibility of producing homogenous thin films with shape memory effect with a further thermal treatment.

ACKNOWLEDGEMENTS

The authors express their gratitude to the CIAM and Conacyt for their financial support with the projects: 107462 and 82515 respectively. Also to M. Morin from the INSA de Lyon, the FIME of the UANL and the CIIDIT for their technical support.TEM work have been made at LABMET, Red de Laboratorios de la Comunidad de Madrid. The help of Isabel Ortiz with TEM is acknowledged.

REFERENCES

1. Yongqing Fu, Sputtering deposited TiNi films: relationship among processing, stress evolution and phase transformation behaviors. Surface and Coatings Technology Vol. 167 (2003) 120–128

2. Y.Q. Fu, Thin film shape memory alloys and microactuators, Int. J. Computational Materials Science and Surface Engineering, Vol. 2, Nos. 3/4, 2009, pp 208-226.

3. K. Otsuka, C.M. Wayman, Shape memory materails, First ed., Cambridge University Press, UK 2002.

4. López-Cuellar Enrique. Fatigue par cyclage thermique sous contrainte de fils à mémoire deforme Ti-Ni-Cu après différents traitements hermomécaniques. Thèse d'Etat, INSA de Lyon, Lyon I, 2002, 180p.

5. De Araujo, C. J. Comportement cyclique de fils en alliage à mémoire de forme Ti-Ni-Cu : analyse electro-thermomécanique, dégradation et fatigue par cyclage thermique sous contrainte. Thèse d'Etat, INSA de Lyon, Lyon I, 1999, 177

6. R. H. Wolf and A. H. Heuer, Journal of Microelectromechanical Systems, 4 (1995) 206.

7. Y.Q. Fu, H. J. Du, W. M. Huang, S. Zhang, M. Hu, Sens. Actuat. 112 (2004) 395-408.

8. S. Miyazaki, A. Ishida, Mater. Sci. Engng. A 273-275 (1999) 106.

9. P. Krulevitch, A. P. Lee, P. B. Ramsey, J. C. Trevino, J. Hamilton, M. A. Northrup, J. MEMS, 5 (1996) 270.

10. J. J. Gill, K. Ho, G. P. Carman, J. MEMS, 11 (2002) 68-77.

11. E. Makino, T. Mitsuya, T. Shibata, Sens. Actuat., 79 (2000) 251-259.

12. Liu HB, Espinosa-Medina MA, Sosa E, et al. Structural Segregation and Ordering of Trimetallic Cu-Ag-Au Nanoclusters. Journal of Computational and Theoretical Nanoscience. (2009) Vol. 6   Issue: 10   Pages: 2224-2227.

13. Lee HM, Mahapatra SK, Anthony JK, et al. Effect of the titanium ion concentration on electrodeposition of nanostructured TiNi films. Journal of Materials Science   (2009) Vol. 44 Issue: 14   Pages: 3731-3735.

14. Troiani H. "Dezincificación y Transformaciones de fase en el sistema Cu-Zn." Tesis (1998)

15. G. De Miccoad, A. E. B., D. M. Pasquevichab . "Caracterización de aleaciones Cu-Zn-Al: Estabilidad térmica de las fases y decincación." Revista Matéria 12: (2007) 245-252

16. Kowalski, M., Spencer, P.J., (1993). "Thermodynamic Reevaluation of the Cu-Zn System." Equi. Diagram, Thermodyn., Calculations 36: 432-438

17. Liang H., Chang Y.A., A Thermodynamic Description for the Al-Cu-Zn System". Equim Diagram, Thermodyn., Calculations 72 (1998) 25-37.

18. Donald L. Smith., Thin-Film Deposition Principles & Practice. Ed. Mc Graw Hill. (1995) 140-145.

19. N. Haberkorn, M. Ahlers and F.C. Lovey., Tuning of the martensitic transformation temperature in Cu–Zn thin films by control of zinc vapor pressure during annealing. Scripta Materialia 61: (2009) 821-824

Mater. Res. Soc. Symp. Proc. Vol. 1276 © 2010 Materials Research Society

Development of an Algorithm for Random Packing of Multi-Sized Spherical Particles

H. de la Garza-Gutiérrez [1,2], G. Plascencia-Barrera[1] and S.D. de la Torre[1]
[1] Instituto Politécnico Nacional. Centro de Investigación e Innovación Tecnológica CIITEC-IPN. Cerrada CECATI S/N. Col. Santa Catarina. Azcapotzalco, C.P.02250, México D.F. Mexico.
[2] Instituto Tecnológico de Chih.-II, Av. de las Industrias 11101, Comp. Ind. Chihuahua, Mexico.

ABSTRACT

A new computational algorithm is introduced for packing simulation of spherical elements/particles into an imaginary container with three main possible geometries, cubic, cylindrical and spherical. The performance of the algorithm depends directly on the strategy or logic considered to solve the problem and the quality of its computational implementation. The combination of these two factors let the packing algorithm here presented and named as Octant Packing Random Algorithm (OPRA) to reduce the computation time between 2 and 127 times, when compared with the simplest or classical Packing Algorithm. OPRA is designed to reduce the number of comparisons needed to accept or reject a new position for an element/particle to be allocated into the virtual container. OPRA considers the container as if it were divided into 8 equal cells or octants limiting the overlap detection for a new position.

INTRODUCTION

In many industrial fields, particles with different size distribution are used. Among these industries are: catalysts, cement, mineral processing, grain storage, etc. In order to accommodate these particles into containers or processing units, it is necessary to understand how such particles are susceptible to utilize the volume destined to their storage/processing. In this paper we propose an algorithm based in random allocation and for this reason, it is useful for systems with low relation solid/volume. To do so, the algorithm takes a particle and transforms it into an element, which can be traced as it accommodates in a given container.

When a system to be modeled involves small particles and/or solid elements placed inside a container, it is convenient to represent both the vessel and the set of elements in a computational arrangement for study. It is then required to specify at least a number of conditions; namely, the size and geometry of the container, the system of used coordinates, the specification of the system's origin and/or spatial location, the geometry of the elements (particles) as well as the number and size of each interacting element, especially when different particle sizes are considered simultaneously.

As for each element and/or particle placed into the vessel concerns, there should be a virtual counterpart allocated inside the system model. The ideal allocation process implies to find out a valid position for each element, where none of it replaces the space occupied by the other. This task is defined to as the packing stage. In order to conduct it, one can opt for developing an original algorithm and/or to choose between some available in the corresponding literature. The packing algorithms can be classified by considering main features of the system, as well as by the products expected. Some criteria to classify them are:

a) Container geometry. Some algorithms are developed for a particular geometry, while others can be easily adapted to work in more than one. The most common geometries used are cubic and cylindrical [12].

b) Element geometry. The selected shape of the elements inside the vessel could be spherical [6,10], cylindrical [5], disc-like [4,7], rings or any other possible geometry. This is an important feature to consider when selecting packing algorithms.

c) Element size. There are systems composed by elements having the same size (mono size), others with a combination of two sizes [4,10] or having more than two sizes (multi-size) [1].

d) Element position. Once a new position for an element has been generated and accepted, some algorithms maintain this position the same during all the process [6]. Other algorithms aiming for a maximum packing density allow the particles position to change by applying different strategies [1, 8].

e) Final density. The selection of a packing algorithm depends completely on the level of compaction to achieve. Not all algorithms are designed to reach high-density level [2,3,6,11].

f) Allocation strategy. Most of the algorithms allocate the particles one by one beginning with the bigger ones, when it is a multi-sized system. Others find out an initial position for all the particles involved in the system, regardless the collisions presented; once this process is done, the algorithm proceeds to resolve the overlaps [1, 4, 12].

g) Partial or complete particles accepted. The container can be visualized as having solid walls, so that only complete particles can be allocated. If a slice of such system is used instead, fraction of particles will rather be considered [1, 6].

One of the easiest algorithms to implement is the Classical Random Packing (CRP). This algorithm takes an element at a time, generate a random position for it inside the virtual container and before accepting such position, it proceeds to detect collisions or overlaps with elements already accepted. If no overlaps are found, the position is accepted and retained. If there are one or more overlaps, the position analyzed is discarded and a new one is generated, repeating the same validation steps. The packing part finishes when all the elements have their own virtual representation.

The Octant Packing Random Algorithm (OPRA) is a variant of the CRP-algorithm, which looks for a reduction in the number of comparisons needed to accept or reject a position for a new element or particle. The OPRA algorithm considers the container as composed by eight equal parts (octants) whereas the overlaps detection is reduced to just few comparisons conducted on the elements previously accepted in only one octant.

Assumptions

A container with cubic geometry and inert particles with spherical shape are assumed to run some tests. A Cartesian coordinate system and only complete elements inside the container are also considered. These assumptions are made to reduce the complexity of the system to be modeled, letting us to be focused on the implementation and performance of the algorithm.

Position Generation

In a Cartesian coordinate system it is necessary to generate 3 random values to define a new position, associating each value to one of the x, y or z axes. These values can be obtained using an internal function of the programming software. The new element position must assure that the element with radius R lies completely inside the container with a length defined by L. In order to assure that only complete elements will fall within the volume of the container, it is necessary to divide the domain into three sub-domains. This can be seen in Figure 1. It is evident that given the geometry conditions of the system, the three sub-domains, are defined as 0 to R, L -2R and L-R. This sub-domains need to fulfill such geometrical restriction, that any given particle must fall within the container; therefore the minimum distance required to do so is that of the radius of the particle. Thus, every single particle can be accounted into the volume, otherwise the particle will be considered out of the system.

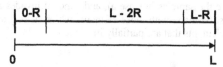

Fig. 1. Range of possible values for a new element with radius R.

Overlap Detection

The overlap taking place between an element already accepted and located in (Xa, Ya, Za) of radius Ra, with a new position to be evaluated (X,Y,Z) for an element with radius R only occurs, if the distance (d) between both particles radii is shorter than the sum of them. The collision between particles then exists and the position comparing process stops, whereas a new searching process begins.

Octant Packing Random Algorithm

In the OPRA algorithm it is considered that the elements/particles container is made out of 8 identical sub-volumes. That is, an imaginary cell composed of 8 octants. The container can be seen as cut by 3 orthogonal planes that cross it through the centre. Thus, it is defined by one plane running parallel to the XY plane, other going parallel to YZ, while the third runs along the XZ plane. This geometric division reduces the number of comparisons required to decide whether a new position is accepted or not, obtaining same results as the Classical Random Packing Algorithm, with the advantage of being faster. Within an octant cell, an element/particle has 3 spatial distribution possibilities (Figure 2). That is; a) either the element lies completely inside the octant, b) the element has its center placed into the octant, although a part of it invades one or more of the neighboring octants and c) elements located in another octant invade it (i.e., reference cell). Cases b) and c) are basically the same, although attention is focused in a different octant.

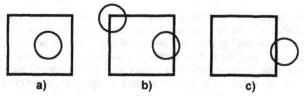

Fig. 2. Three possibilities for an element/particle to fit into an octant cell.
a) completely inside, b) partially out and c) partially inside.

41

The algorithm function for overlap detection must take those 3 cases into account. After a new position XYZ is randomly generated, it should identify into which octant it is located. Subsequently, through the usage of the radius-R element its particular spatial situation is determined. Case a) implies a fully contained particle into an octant and case b) partially out of limits. That is, if the new element would be allocated in the designed position it may or not invade other octants. When the whole element/particle is fully contained within an octant, the conducting overlap detection algorithm should involve commands as to be comparing with the location of those elements belonging to its own octant and with those elements as in case c) that are invaders from others octants. If the new element is partially out of limits, Case b), the matching-comparisons are the same as in Case a), and, since it invades at least one other octant, overlap comparisons must be made with those elements which belong to the invaded octant and as it does, with all those elements that are partially inside it.

Implementation

Once a new position for an element/particle of radius R is generated, its octant must be identified. One way to do that is determining whether the position XYZ is set, either to the left or right, at front or behind and/or up or down, by considering the center of the cubic container as point of reference. Thus, octant number 1 results when the element is placed to the left, at the front and in the upper most side, considering the particular octant numbering as appears in Figure 3. The process continues identifying if the element is in a situation like case a), completely contained into one octant, or case b) invading one or more octants.

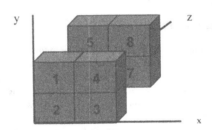

Fig. 3. Cubic volume divided into eight equivalent cells/parts.

Accepting an element/particle within an octant

It is necessary to identify whether a particle invades or not other octants. Having this considered, algorithm will execute the overlap detection according to the particular situation of the element. When overlapping does not detect any collision, the position of the element is accepted. The elements position is then registered at each octant, recording the X, Y, Z and R-values for each of these elements. This data is later used for comparison purposes, when a new element is placed in the volume. The same steps must be repeated until all elements are allocated without overlaps.

Adaptation to cylindrical and spherical container geometries

OPRA-algorithm can easily be adapted to perform well under other basic geometries, such as cylindrical and spherical. To adapt OPRA to a cylindrical geometry it is necessary to choose between some spatial possibilities. The coordinate system selected can be located at the center of the top, at the center of the base or at the center of the cylinder. The position for an element can be generated under cylindrical coordinates at the beginning of the programming setting. Caution, however, should be taken as to convert them into the Cartesian system. The later is in order to calculate the distance between elements during the overlapping detection. In case to generate the new position under a Cartesian coordinate system, the obtained values must first be evaluated to be sure that the element does not go further the limits of the cylinder. If it occurs, this position may be discarded and a new one must be generated, or it might be convenient to adjust the position following direction of the hypotenuse as to make it fitting inside the limits of the cylinder.

For a spherical geometry container the origin of coordinates can be set at the center of the sphere. Although the elements positions should be established since the beginning of any experimental run using spherical coordinates, the system must be converted into Cartesians equivalent for the overlapping detection. If this is selected to generate Cartesian positions however, care must be taken because the XYZ generated values define a cube, where the sphere is placed inside, existing regions in the cube that are not part of the sphere but elements which go further the sphere limits.

RESULTS AND DISCUSSION

In order to evaluate the performance of the OPRA-algorithm different data sets are used. These included mono-size, binary size and multi-sized systems. The CRP-algorithm is also used for performance comparison. For testing the OPRA-algorithm, it is selected a cubic cell container of L = 100 µm and a spherical assembly for the elements. The data sets used are taken from a Particle Size Distribution (PSD) analysis of an Ultra Fine Portland Cement material, which its hydration mechanism is currently being modeled. The reported processing times are obtained using a MacBook Pro computer with CoreDuo processor and are dependent of the computer capacity used, in this case. Useful criteria to compare the performance disclosed from both algorithms, which is independent of the computer type, is obtained by dividing the time required by the CRP and the time required by the OPRA. This relation is analyzed and is shown as column in Tables I-II and as row in Tables III-V.

Mono-size System

Three different radii are used for the studied elements; namely, 8.75 µm, 3.1 µm and 1.9 µm, in combination with 3 elements in the container. Comparing the computer time required by each algorithm, it is clear that the Octant Packing Random Algorithm works faster than its Random counterpart. According to the results obtained for the mono-size system, the OPRA performance is at least four times faster, where this efficiency is directly related to the volume fraction of solids. The Volume of Solids column shows the ratio between the volumes occupied

43

by all particles to the container volume. Thus, the maximal volume of solids expected to achieve when using both algorithms for a mono-size system is about 0.33 (33%).

In the first run trial, the radius used is 8.75 μm, having a relation of D/L = 0.175, which is the diameter of the particle (D) divided by the size of the container (L). This system of large particles shows a wide range in the time relation for both algorithms, being 4.8 times faster the OPRA than the CRP for a loose density system (composed of 100 particles), and up to 128 times faster, when the volume of solids is close to the maximum limit for a mono-size system (110-111 particles). For runs performed using a radius of 3.1 μm and D/L = 0.062, a volume fraction of solids of 0.33 can be reached. In these runs the CRP/OPRA time relation goes from 6.61 to 10.1 times faster the OPRA than the CRP. The difference between the performances of both algorithms rises while increasing the number of particles to allocate and also when the volume of solids is close to the limit for such mono-size system. Using a particle radius of 1.9 μm and D/L = 0.019, runs are made using three different numbers of particles to allocate (7500, 10500 and 11000). The CRP/OPRA time relation for these three runs remained around 7, showing a light increase when the number of particles is higher (see Table I).

Table I. Mono-size system using three different radii.

Radius μm	Number of Elements	Fraction of Solids	Time (sec) CRA Classical Random	Time (sec) OPRA Octant Packing	CRA/ OPRA
8.75	100	0.281	1.281	0.265	4.83
8.75	110	0.308	87.66	1.83	47.92
8.75	111	0.311	354.52	2.77	127.98
3.1	2,300	0.287	26.86	4.06	6.61
3.1	2,500	0.312	97.641	13.97	6.99
3.1	2,650	0.331	534.02	52.89	10.10
1.9	7,500	0.215	35.20	5.03	7.00
1.9	10,500	0.302	649.16	86.45	7.40
1.9	11,000	0.316	1443.37	193.42	7.46

Binary-size System

Computational runs are conducted using four radii combinations (8.75, 4.3), (8.75, 1.9), (6.3, 4.3) and (6.3, 1.9). The number of large particles used is set close to its maximum value (i.e., 105 particles of 8.75 μm and 300 particles of 6.3μm), letting the small ones to accommodate into the free space. The CRP/OPRA ratio does not establish neither lower nor upper limit; but it shows the performance of both algorithms for the specific combination of radii and the number of particles of each size used. Thus, the OPRA algorithm is 87.5 times faster than the CRP for the combination of 300 particles with radius of 6.3 μm with 350 particles of 4.3 μm. Two other computational runs carried out with smaller radius (1.9 μm), as a second group of particles lead the system to reach a solids volume fraction of 0.49 whereas the CRP/OPRA time ratio revealed slight variation. For the runs having 8.75 μm and 1.9 μm radii, the time relation went from 5.08 to 8.74 and from 3.38 to 7.9 for the runs with 6.3 μm and 1.9 μm (Table II).

Table II. Binary-size system using four different combinations of two radii.

Rad. R1	Rad. R2	Elements with R1	Elements with R2	Frac. of solids	Time (sec) Classical R.	Time (sec) Octant Pack.	CRA/ OPRA
8.75	4.3	105	400	0.428	17.48	1.25	13.98
8.75	4.3	105	470	0.451	275.22	9.92	27.74
8.75	4.3	105	480	0.454	1008.52	18.125	55.64
8.75	1.9	105	6,000	0.467	153.66	30.22	5.08
8.75	1.9	105	6,500	0.481	440.28	72.66	6.06
8.75	1.9	105	6,800	0.490	1110.87	127.05	8.74
6.25	4.3	300	290	0.403	31.37	4.3	7.3
6.25	4.3	300	300	0.407	51.81	5.09	10.17
6.25	4.3	300	320	0.413	177.61	8.61	20.63
6.25	4.3	300	350	0.423	2750.78	31.343	87.76
6.25	1.9	300	5000	0.450	71.45	21.08	3.38
6.25	1.9	300	5600	0.468	213.11	57.07	3.73
6.25	1.9	300	6000	0.479	585.12	123.81	4.72
6.25	1.9	300	6200	0.488	1563.06	268.34	5.82
6.25	1.9	300	6500	0.494	3725.34	470.92	7.9

Multi-size System

As for the multi-size system, eight and nine radii values are used while determining the number of elements required for each section. The later according to the particle size distribution data obtained from an Ultra-fine Portland Cement. Table III shows computational runs using 8 different sizes while changing the amount of small particles for each run (radius 1.9 μm). The time relation between the studied algorithms went from 2.94 to 8.29 times faster the OPRA than the CRP. It is clear again that for such systems with low volume fraction of solids, the behavior of both algorithms is closer. The solids volume fraction reached in these runs are from 0.487 to 0.53.

Table III. Multi-size system using 8 radii and varying the number of particles with radius 1.9 μm.

Radii μm	Num Part	Num Part	Num Part	Num Part
0.65	0	0	0	0
1.9	3494	4000	4500	5000
3.1	404	404	404	404
4.3	151	151	151	151
6.25	49	49	49	49
8.75	19	19	19	19
12.0	8	8	8	8
15.5	3	3	3	3
21.0	2	2	2	2
31.0	0	0	0	0
Time CRP (sec)	20.14	45.83	146.26	729.34
Time OPRA (sec)	6.78	14.14	33.42	87.92
CRP/OPRA	2.97	3.24	4.37	8.29
Fraction of solids	0.487	0.501	0.516	0.530

Computer results obtained for three additional runs using more amount of material are listed in Table IV. In the first series group, each section identified by its radius, the material amount is proportionally increased, so that the number of particles grew in each case. Considering these values as a starting point, during second and third runs, the amount of particles for the smallest size is increased. The volume fraction of solids went from 0.551 to 0.568, and the time relation values varied from 6.09 to 25.33. It is interesting to note that for the second run, the CRP required 2,407 seconds (40 min) versus 151 seconds (2.51 min) used by OPRA, with a CRP/OPRA time relation of 15.92. For the third run reported in Table IV, the CRP required 5,101.5 sec (85 min) while the OPRA needed 201.4 seconds (3.35 min) with a time relation of 25.33. Evidently, the OPRA algorithm had a significant performance at those conditions.

Table IV. Multi-size system increasing the amount of total material.

Radii µm	Num Part	Num Part	Num Part	Num Part	Num Part
0.65	0	0	0	0	0
1.9	3994	4500	4600	2800	2900
3.1	461	461	461	560	560
4.3	172	172	172	211	211
6.25	56	56	56	68	68
8.75	21	21	21	27	27
12.0	9	9	9	9	9
15.5	4	4	4	5	5
21.0	2	2	2	3	3
31.0	0	0	0	0	0
Time CRP (sec)	259.83	2407.25	5101.53	2303.25	5688.89
Time OPRA (sec)	42.67	151.17	201.42	101.66	143.87
CRP/OPRA	6.09	15.92	25.33	22.65	39.54
Fraction of solids	0.551	0.565	0.568	0.625	0.628

Table V. Multi-size system using 9 radii and varying the amount of material in each run.

Radii µm	Num Part	Num Part	Num Part	Num Part	Num Part	Num Part	Num Part
0.65	43655	46541	47938	49847	52671	55811	60630
1.9	1747	1866	1921	1997	2111	2235	2429
3.1	404	430	442	461	486	515	560
4.3	151	161	166	172	183	194	211
6.25	49	53	56	56	60	64	68
8.75	19	21	20	21	22	24	27
12.0	8	7	8	9	8	9	9
15.5	3	4	4	4	4	4	5
21.0	2	2	2	2	3	3	3
31.0	0	0	0	0	0	0	0
Time CRP (sec)	486.72	613.22	688.37	815.70	1194.58	1877.94	5807.98
Time OPRA (sec)	154.95	212.37	250.61	309.14	557.23	933.58	2525.37
CRP/OPRA	3.14	2.88	2.74	2.63	2.14	2.01	2.30
Fraction of solids	0.486	0.518	0.532	0.551	0.603	0.634	0.684

Additional runs (Table V) are performed including a 10% of fine material having radius of 0.65 μm and a D/L = 0.013. These runs included a great number of very fine elements (from 43,000 to 60,000) leading the time relation between both algorithms to its lower value, i.e., 2.01 times faster the OPRA than the CRP. However, a reduction of 50% in the elapsed time becomes significant if we consider that the saved time is 20 or more minutes. In the last run of this experimental set the CRP/OPRA time relation start to rise, up to 2.30.

CONCLUSIONS

In many industrial and modeling processes it is necessary to understand how powder particles (either metallic, ceramic, or plastic), having various size distribution fit into a given volume. This information is useful for enhancing sintering and/or densification operations of materials. A new computational algorithm named Octant Packing Random Algorithm (OPRA) is introduced here for packing simulation of spherical elements/particles into an imaginary container with three main possible geometries, cubic, cylindrical and spherical. OPRA does the same job as the simply Random Packing Algorithm (CRP), but it works 2 to 128 times faster depending basically on the number of particles to allocate and the kind of system (mono-size, binary size or multi-size) used.

The OPRA maintains the position of each particle fixed during all the allocation process, so the OPRA does not use the volume fraction in its best, thereby high densities cannot be expected. For mono-sized systems the OPRA performance goes from 4 to 128 times faster than the CRP, and this efficiency is directly related to the volume fraction of solids to reach. As for mono-sized systems it can be expected a maximum volume fraction of solids of about 0.33.

As for binary systems, the OPRA is 87 times faster when the two radii (CRP/OPRA) are similar (8.75,4.3) or (6.25,4.3). However, the performance between both algorithms is closer if the radii of one of the particles considered is small in comparison with the other (8.75, 1.9) or (6.25, 1.9). The volume fraction of solids achieved in the binary systems studied is close to 0.50. The maximal volume fraction of solids in the multi-size systems with 8 radii, is 0.568, and the time relation between both algorithms took values up to 25.33. The OPRA had a very significant performance for these conditions.

The computational runs that included a great number of very fine elements using 9 different radii, lead to a time relation between both algorithms to a value of 2.01 times faster the OPRA than the CRP. However, a reduction of 50% in the elapsed time is significant. The maximum volume fraction of solids reached considering all runs is 0.684, which is an important contribution in this particles packing-field.

ACKNOWLEDGEMENTS
HGG acknowledges to Conacyt-Mexico for his doctoral grant. This work was partially supported by Instituto de Ciencia y Tecnología del Distrito Federal ICyTDF, through the project PIFUTP08-110. Authors also acknowledge to SNI-Conacyt.

REFERENCES

[1] Stroeven P, Stroeven M., Image Anal Stereol 19: 85-90 and 22:1-10 (2003)

[2] Sobolev K, Amijanov A., Advanced Powder Tech. 15(3), 365 (2004)

[3] Sobolev K, Amirjanov A., Powder Tech. 141, 155 (2004).

[4] Okubo T, Odagaki T., J. of Physics: Condensed Matter. 37, 6651 (2004).

[5] Sugihara K., J. of the Japan Society for Simulation Tech. 24(2),166 (2005).

[6] Nolan G.T., Kavanagh P.E., Powder Tech. 72(2), 149 (1992).

[7] Uche Q.U., Stillinger F.H., Torquato S., Physica A: Statical Machanics and its Applications, 342(3-4), 428 (2004).

[8] Cheol K. J., Martin M.D, Sung L.Ch., Powder Tech., 123(3), 211 (2002).

[9] Nolan G.T., Kavanagh P.E., Powder Tech., 84(3), 199 (1992).

[10] Vrabecz A., Tóth G., Molecular Physics. 104(12), 1843 (2006).

[11] Santiso E., Muller E.A., Molecular Physics. 100(15), 2461 (2002).

[12] Abreu C.R.A, Macias-S. R, Tavares F.W., Castier M. Braz. J. Chem. Eng., 16(4), (1999).

Mater. Res. Soc. Symp. Proc. Vol. 1276 © 2010 Materials Research Society

Development of a New Nickel-base Superalloy for High Temperature Applications

Octavio Covarrubias[1]

[1]Frisa Aerospace SA de CV, Valentin G Rivero 200, Col. Los Treviño, Santa Catarina, Nuevo León, 66150, México. E-mail: ocovarrubias@frisa.com

Keywords: 718Plus®, forging, heat-treatment.

ABSTRACT

Since their appearance during in the 1940 decade, nickel-base alloys are appreciated for their superior mechanical properties and microstructural stability at elevated temperatures and high stresses. They are typically used in jet-engines and land-based turbines for energy generation. Such materials, known as *superalloys* are in constant evolution as designers are encouraged to propose more efficient and powerful systems of propulsion and energy generation. This evolution leads to conceive and manufacture new superalloys capable to fulfill higher requirements. Alloy 718Plus® is emerging as an alternative material for the design and construction of components to be used in jet-engines and land-based turbines for energy generation. 718Plus® is a precipitation hardened nickel-base alloy designed to have the stability of superalloys similar to Waspaloy and the good processing characteristics of other materials as the 718 alloy. Since the early 2000 decade, ATI Allvac has lead a complete program in order to validate capabilities and properties of the 718Plus® alloy. Objectives for this effort include a characterization and its introduction as a viable material for the design and manufacture of components to be installed technologically. As part of this project, contoured rings made of 718Plus® are rolled considering industrial conditions. Several heat treatments, involving solution and precipitation processes are performed on segments extracted from involved contoured rings. Effects of such hot-working conditions and heat treatment procedures on properties as forgeability, tensile, hardness and stress-rupture characteristics are evaluated. Optical and electron microscopy are performed to evaluate microstructural properties as grain size and promotion of precipitates, in order to complement reported results.

INTRODUCTION

Performance improving of jet-engines and land-based turbines for energy generation is a constant for nowadays designers. Use of better and more capable materials to reach this objective is an important strategy, where superalloys have a primordial role. Superalloy ATI 718Plus® is emerging as a viable material for the manufacture of components to be used in jet-engines and land-based turbines. This material is a precipitation hardened nickel-base alloy designed to have excellent stability and metallurgical properties at elevated temperature in combination with good processing properties [1].

ATI Allvac is leading a complete program, joining research centers, manufacturing operations and original equipment manufacturers (OEM's), to characterize and validate properties and capabilities of ATI 718Plus® in order to consider this superalloy as an appropriate material for the design and manufacture of components to be installed in jet-engines and land-based turbines for energy generation [2]. Some of these components can be produced by ring-rolling processing, which is a manufacture option that promotes important metallurgical benefits as grain size control and superior mechanical characteristics.

Ring-rolling operations include use of specialized tooling to produce contoured shapes. Contoured rings can provide complex shapes and its implementation leads to a reduction of input weight, improving of machining touch times and, in some cases, elimination of welding operations. Manufacturing parameters such as temperature, deformation ratios and deformation rates are determined and controlled to have a robust and reliable process. Considering the processing of most superalloys, such parameters have important effects in ATI 718Plus® microstructure and mechanical properties. A better understanding of ATI 718Plus® behavior will give designers key information about its capabilities and possible applications. Extensive research is done in this field [3, 4, 5, 6, 7]. This report is a contribution to these efforts.

EXPERIMENT

A VIM+VAR 203.0 mm diameter billet is selected to extract several segments for the production of contoured rings with a weight of 80.0 kg and dimensions: 560.0 mm outer diameter, 455.0 mm internal diameter and 150.0 mm height. Table I summarizes the chemical composition of the starting material.

Table I: Chemical Composition of 718Plus® Starting Material

Chemistry	C	Cr	Mo	W	Co	Fe	Nb	Ti	Al	Ni
% w/w	0.020	17.88	2.69	1.03	9.15	9.72	5.48	0.76	1.51	Bal.

Extracted segments are forged and rolled considering production parameters. Involved temperatures for hot-working operations are between 950°C and 1200°C. Contoured rings are cut and segments exposed to the heat treatment conditions listed in Table II. Some of such segments are ring slices used for grain size mapping after forging operations and after each heat treatment operation.

Table II: Heat Treatment Conditions for ATI 718Plus® Ring Segments.

Condition	Solution heat treatment	Precipitation heat treatment
HT1	950°C/2hr/Air-cooling	788°C/8hr/furnace cooling to 704°C + 704°C/8hr/Air-cooling
HT2	980°C/2hr/Air-cooling	788°C/8hr/furnace cooling to 704°C + 704°C/8hr/Air-cooling
HT3	1010°C/2hr/Air-cooling	788°C/8hr/furnace cooling to 704°C + 704°C/8hr/Air-cooling

Samples from the "as forged" condition and from each heat treatment condition are evaluated by optical microscopy in order to determine grain size evolution. Characterization of samples, relative to the presence of precipitates, is evaluated by scanning electron microscopy.

Mechanical testing including tensile tests at room and elevated temperature, hardness and stress-rupture, is performed on heat treated samples per specifications AMS 5441 and AMS 5442. Elevated tensile testing is performed at 704°C. Stress-rupture testing is performed at 704°C, initial load of 621 MPa; after 39.0 hours of testing incremental loading is performed with increments of 34.5 MPa at intervals of 8.0 hours.

RESULTS AND DISCUSSION

Evaluation of samples after hot-working operations, or "as forged" condition, reveals a uniform fully re-crystallized microstructure. The microstructure is a consequence of the forging/rolling parameters considered for ring manufacture. Reported average grain size is 44.9 μm (ASTM 6.0). A uniform microstructure along the transversal section of each segment is observed, despite their contoured shape.

Samples exposed to condition HT1 show an average grain size of 37.8 μm (ASTM 6.5). Limited or null grain coarsening is evident after solution treatment at 950°C. When ATI 718Plus® segments are exposed to condition HT2 with a solution temperature of 980°C, some grain size coarsening is observed. For this condition, average grain size is reported as 53.4 μm (ASTM 5.5). For those segments exposed to heat treatment conditions of HT3, where a solution temperature of 1010°C is considered, grain coarsening is evident: reported average grain size is 63.5 μm (ASTM 5.0). Figure 1 illustrates the observed microstructures.

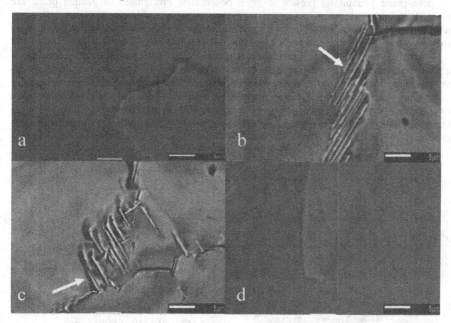

Figure 1: Representative microstructures of ATI 718Plus® alloy: a) As-forged condition, (b) HT1 heat treatment (arrow indicates delta-phase precipitates at grain boundary), (c) Microstructure after HT2 heat treatment (arrow indicates limited delta-phase precipitates at grain boundary). (d) Microstructure of alloy after HT3 heat treatment, no evident delta-phase precipitates at grain boundaries are seen.

51

A lack of delta-phase precipitates is observed by electron microscopy for the "as forged" sample condition. When the rolled sample material is exposed to condition HT1, delta precipitates are evident at the grain boundaries. As rolled samples exposed to heat treatment HT2 develop delta-phase at grain boundaries; their quantity and size is reduced when compared with samples exposed to HT1 condition. Samples exposed to condition HT3 develop a microstructure free of evident delta precipitates at grain boundaries, since the alloy is solution treated above the delta solvus temperature. Another characteristic of HT3 is a smooth texture on grain boundaries.

The effect of the solution condition is reflected by the tensile properties of ATI 718Plus®. A drop on tensile properties is observed as a function of increasing solution temperature: as samples are exposed to 950°C, 980°C and 1010°C temperatures for HT1, HT2 and HT3 heat treatment conditions, lower tensile properties are promoted at room and elevated temperatures. All samples exposed to 1010°C during solution process of HT3 condition failed yield strength requirements at room and elevated temperature testing.

An adequate microstructure for ATI 718Plus® alloy must consist of recrystallized grains with delta-phase precipitates present at grain boundaries. Precipitates of gamma prime are promoted during precipitation procedures and they are responsible for alloy strengthening. During solution operations, delta-phase precipitates at grain boundaries and have a more important effect on stress rupture properties, which will be discussed later.

Tensile results at room temperature reveal the effect of solution temperature for each of the three heat treatment procedures: as the solution temperature is increased, tensile properties decrease due to coarsening of the average grain size and inhibition of delta phase grain boundary precipitation. Additionally the ductility (reduction of area, R/A) decreases at room temperature as solution temperature is increased. This behavior can be related to limited formation of delta-phase precipitates at grain boundaries. Figure 2 summarizes yield strength (YS) results at room temperature.

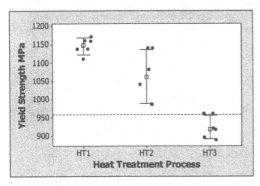

Figure 2: ATI 718Plus® YS properties at room temperature after solution and precipitation procedures. Minimum requirement per AMS 5441 and AMS 5442 is 958 MPa (dashed line).

The tensile properties at 704 °C show a similar behavior. They tend to decrease as the alloy is exposed to the higher solution temperatures involved by the different heat treatment processes. Figure 3 presents the yield strength as a function of heat treatment at 704 °C. As for ductility, the reduction of area (R/A) properties at elevated temperature are consistent, but with an inverse slope tendency to that reported for the room temperature testing.

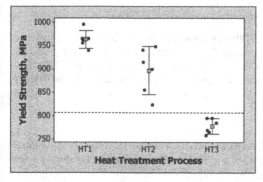

Figure 3: ATI 718Plus® YS properties at 704°C after solution and precipitation procedures. Minimum requirement per AMS 5441 and AMS 5442 is 807 MPa (dashed line).

Alloy ATI 718Plus® shows a stress-rupture behavior similar to other nickel-base superalloys. Samples extracted from prototypes exposed to heat treatment HT1 present a finer average grain-size and an important promotion of delta-phase precipitates at grain boundaries, developing better stress-rupture properties. On the other hand, when samples of alloy ATI 718Plus® are treated according to the HT2 procedure, it is found that grains coarsen and a limited formation of delta-phase precipitates at grain boundaries as a result of higher solution temperature. For this condition, two failures are found suggesting that the described microstructural characteristics limit stress-rupture properties.

Stress-rupture samples extracted from ATI 718Plus® exposed to HT3 heat treatment conditions present the coarsest average grain-size and an absence of delta-phase precipitates at grain boundaries. All samples with this microstructure fail stress-rupture tests, confirming the importance of a fine average grain-size and the presence of delta-phase precipitates at grain boundaries for 718Plus® alloy when it is evaluated per AMS 5441 or AMS 5442. Figure 4 shows time to rupture results for the performed stress-rupture experiments.

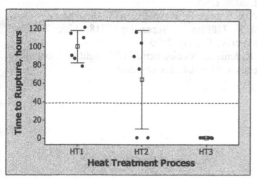

Figure 4: 718Plus® stress-rupture properties at 704°C and incremental loading after solution and precipitation procedures. Requirement per AMS 5441 and AMS 5442 is 39.0 hours as minimum time to rupture (dashed line).

CONCLUSIONS

Production of contoured rings with ATI 718Plus® alloy for aerospace and energy generation applications is feasible. For the conditions evaluated, a good balance of mechanical properties can be obtained with an average grain size of 37.8 micron (ASTM 6.5) for ATI Allvac 718Plus®. For the heat treatments evaluated, delta-phase at grain boundaries can be heavily promoted when alloy is exposed to 950°C

ACKNOWLEDGEMENTS

Frisa Aerospace SA de CV recognize Conacyt-Mexico support by INNOVAPYME project 110988 and contribution of R. Jeniski, S. Cox and E.T. McDevitt from ATI Allvac for the realization of this research.

REFERENCES

1. ATI 718 Plus® Alloy Data Sourcebook, Version 1.1, Allegheny Technologies Company, 2010, USA.
2. E.A. Ott, J. Groh & H. Sizek, Metals Affordability Initiative: Application of Allvac Alloy 718Plus® for Aircraft Engine Static Structural Components, Superalloys 718, 625, 706 and Derivatives 2005, The Minerals, Metals & Materials Society, 2005, USA.
3. X. Xie, C. Xu, G. Wang, J. Dong, W.D. Cao & R. Kennedy, TTT Diagram of a Newly Developed Nickel-Base Superalloy- Allvac® 718Plus®, Superalloys 718, 625, 706 and Derivatives 2005, The Minerals, Metals & Materials Society, 2005, USA.
4. R.A. Jeniski & R.L. Kennedy, Development of ATI Allvac® 718Plus® Alloy and Applications 718, 625, 706 and Derivatives 2005, The Minerals, Metals & Materials Society, 2005, USA.
5. W.D. Cao & R.L. Kennedy, Recommendations for Heat Treating Allvac® 718Plus® Alloy Parts, ATI Allvac Research & Development, 2006, USA.
6. X. Xie, C. Xu, G. Wang, J. Dong, W.D. Cao & R. Kennedy, TTT Diagram of a New Developed Nickel-Base Superalloy Allvac® 718Plus™, Superalloys 718, 625, 706 and Derivatives 2005, The Minerals, Metals & Materials Society, 2005.
7. R.L. Kennedy, W.D. Cao, T.D. Bayha & R. Jeniski, Developments in Wrought Nb Containing Superalloys (718+ 100°F), The Mineral, Metals & Materials Society, 2003.

Mater. Res. Soc. Symp. Proc. Vol. 1276 © 2010 Materials Research Society

Stress Ratio Effect on Fatigue Behavior of Aircraft Aluminum Alloy 2024 T351

M. Benachour[1], A. Hadjoui[1], M. Benguediab[2], N. Benachour[3]

[1] Automatic Laboratory of Tlemcen, Mechanical Engineering Dpt, University of Tlemcen, BP 230, Tlemcen, 13000, Algeria.

[2] Physical Mechanics and Materials Laboratory, Mechanical Engineering Dpt, University of Sidi Bel Abbes, 22000, Algeria

[3] Department of Physics, University of Tlemcen, 13000, Algeria.

ABSTRACT

Aluminum alloy series 2xxx, 6xxx, 7xxxx and 8xxx enjoy the widest use in aircraft structural applications. Among these materials, aluminum alloy 2024 remains the most commonly used and especially in T351 temper situation. The fatigue crack propagation behaviour of aluminum alloy 2024 T351 has been investigated using V-notch specimen in four bending test. A series of stress ratios from 0.10 to 0.50 was investigated in order to observe the influence of stress ratio on the fatigue life and fatigue crack growth rate (FCGR). The increase in FCGR, which occurs as the stress ratio is increased from 0.10 to 0.50, is generally attributed to an extrinsic crack opening effect. In T-S orientation and at low stress intensity factor, the increasing of stress ratio increase the FCG. Experimental results are presented by Paris law when coefficients C and m are affected by stress ratio. Contrary, at high stress intensity factor, the effect of stress ratio is reversed. We notice a decreasing of fatigue crack growth rate with an increasing of stress ratio. This effect may be explained by microstructure effect in (T-S) crack growth. The analysis of stress ratio effect by Elber model, shown that this model gives bad interpolation in this situation and the parameter characterized the crack closure factor will be adjusted.

INTRODUCTION

The problem of fatigue behavior of materials in mechanical structures, machine parts, etc. is a crucial point in predicting the fatigue life. In general, the fatigue process is depicted by three distinct regions. Region I is associated with the growth of cracks with low ΔK_{th}, and is commonly believed to account for a significant proportion of the fatigue life of a structure. Region II has received the greatest attention as it is in this region where the "Paris" crack growth law "Paris" [1] can be applied. Several different variants of the Paris crack growth law have evolved by many researchers [2-4]. Finally, region III is associated with rapid crack growth. Predicting the fatigue crack growth rate at constant, variable or random loading is of practical interest for many aeronautical applications, aerospace, automobile, etc. A major concern of fracture mechanics is the influence of the stress ratio on the behaviour of cracks, which is classically defined as: the ratio of minimum to maximum applied stresses. Many empirical models for fatigue crack growth have been proposed in the literature to account for the stress ratio dependence of FCG curves (Forman model [5]; Walker model [6]). It was argued that the reason for this influence is the crack closure effect, which introduced first by Elber [7].

Crack growth orientation is an others parameter who affects the fatigue crack growth. Sinha et al [8] attributed the differences observed between the fatigue crack growth rate for two load ratio (R=0.1 and R=0.8) in both directions T-L and L-T for the alloy Ti-6Al-4V unlike the level closure. The investigation of Hariprasad et al. [9] have shown that the crack growth resistance was affected and important for L-T, L-S and T-S orientation comparatively to the T-L, S-T et S-L orientation for aluminum alloy (Al-8.5 pct Fe-1.2 pct V-1.7 pct Si). Under different stress ratio, Kermanidis et al. [10] have show a shift of crack growh curves for aluminium alloy 7475 T7351 and 2024 T851. The resulting effect of an increase of stress R on FCGR da/dN has been investigated for different materials [11-14]. Generally, an increase in R results in an increase in da/dN for a given stress intensity range, DK. This influence of R can essentially come from two sources: a true material dependence of crack growth rate on R (an intrinsic material effect) and/or a crack closure effect. Recently, in the investigation of Lee et al [15], the FCGR (da/dN) increases and the threshold ΔKth decreases with increasing stress ratio R under constant amplitude loading in different environment

The present work intends to analyse the fatigue crack propagation in heat-treated aluminium alloys. For this purpose, fatigue crack propagation tests have been performed in 2024-T351 aluminium alloy. The influence of stress ratio in T-S orientation is analysed.

EXPERIMENTAL PROCEDURE

Material and specimen geometry

The experimentation was performed on the Al 2024 alloy, widely used for aeronautical applications, supplied in the form of 30 mm thick rolled plate. Aluminum 2024 alloy received the T351 thermomechanical treatment (hardened and tempered). The chemical composition of studied material is listed in Table I. The mechanical properties at room temperature are shown in Table 2.

Table I. Chemical composition of Al 2024 T351 (wt%)

Si	Fe	Cu	Mn	Mg	Cr	Zn	Ti	Ni	Pb	Al
0.105	0.159	3.97	0.449	1.50	0.05	0.109	0.018	0.02	0.056	Bal

Table II. Mechanical properties of Aluminum alloy 2024 T351

E (GPa)	$\sigma_{0.2}$ (MPa)	UTS (MPa)	A(%)
74	363	477	12.5

The fatigue tests are performed on V-notch specimens with 45° angle in four bending. The fatigue specimen and dimension are shown in figure 1. Specimens are polished to 10 μm finish on the surfaces crack growth. Specimens have (b x w = 10 x 10 mm^2) section with an initial length a_0 for the notch.

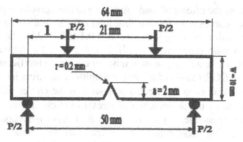

Figure 1. V-notch specimen in four points bending test

Fatigue crack growth measurement

Fatigue crack growth tests were performed using closed-loop servo-hydraulic testing machine "MTS 810" with 100 KN load capacity under applied constant amplitude sinusoidal wave loading at the frequency of 10 Hz and R= 0.1. Specimens are subjected to the bending fatigue tests. Stress intensity factor for V-notch bent specimen is expressed by the following expression [16]:

$$K = \frac{3Pl\sqrt{\pi a}}{bw^2} . f(a/w) \tag{1}$$

where f(a/w) is the geometry correction function given by:

$$f(a/w) = 1.122 - 1.4(a/w) + 7.33(a/w)^2 - 13.08(a/w)^3 + 14(a/w)^4 \tag{2}$$

In fatigue tests, we note the length of the crack (mm) versus number of cycles N (a=f(N)). Secant method was used for modeling fatigue crack growth rates.

RESULTS & DISCUSSION

The fatigue cracking has undergone at constant amplitude for load ratio range (0.1≤R≤0.5) at a frequency of 10 Hz. Table III gives fatigue life (number of cycles) corresponding to a crack length between an initial value "a_0" and a final value "a_f" for different loading levels and same load ratio. The initial crack length was obtained by pre-cracking. Also the initial and final values of stress intensity factor are given. The fatigue crack propagation is expressed by the Paris law [1] where the parameters C and m, material characteristics are determined

Table III. Fatigue life for different load ratio

R	a_0 (mm)	a_f (mm)	P_{max} (KN)	N (cycles)	ΔK_0 (MPa√m)	ΔK_f (MPa√m)
0.1	3.34	7.875	1.149	382 000	5.395	22.69
0.2	3.31	7.140	1.184	569 700	4.85	16.88
0.3	3.365	7.365	1.160	547 000	4.22	15.90
0.5	2.735	6.280	2.500	240 000	5.535	15,82

In order to show the effects of load ratio, fatigue life curves are plotted according to crack length. Figure 2 gives curves a = f (N) for four load ratio R. We note that for the same load, the fatigue life increases with increasing load ratio R = 0.1 to R = 0.3. The evolution of these results is in good agreement with others works [17]. For R = 0.5, the fatigue life is affected by the amplitude of loading, where a decrease in number of cycles is significant.

Evolutions of FCGR (da/dN) function of amplitude of stress intensity factor are plotted in figure 3. The effect of load ratio is reflected by a shift of the curves da/dN-ΔK to the low values of ΔK. These results show the dependence of threshold stress intensity factor ΔK_{th} on the load ratio. Unlike, for high stress intensity factor, the shift of curve is reverse. We notice a decreasing in FCGR with increasing load ratio. This effect can be explained by the effect of the microstructure in the direction of propagation (T-S). The solid lines represent the approximation by Paris Law. The parameters of the Paris law are affected by load ratio (Table IV). We note a decreasing of the coefficient "m" when the load ratio increases.

The evolution of FCGR (da/dN) as a function of effective stress intensity factor range ΔK_{eff} is shown in figure 4. The effective stress intensity factor is evaluated using Elber model (U=0.5 + 0.4R) for different load ratios. The results show a good calibration of cracking curves for low stress intensity factor. Although there have been bad timing curves cracking for high stress intensity factor. For this reason, coefficients of Elber model must be adjusted to have an intrinsic curve. For R = 0.5 no closure have be show.

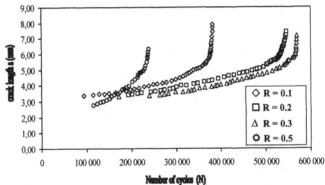

Figure 2. Effect of load ratio on fatigue life for Aluminum alloy 2024 T351

Table IV. Coefficient of Paris law for da/dN (mm/cycle)

R	C	m
0.1	7.0×10^{-10}	4.89
0.2	1.0×10^{-9}	4.60
0.3	2.0×10^{-8}	3.41
0.5	8.0×10^{-8}	2.83

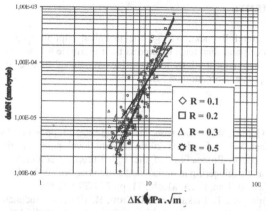

Figure 3. Effect of load ratio on FCGR for Aluminum alloy 2024 T351

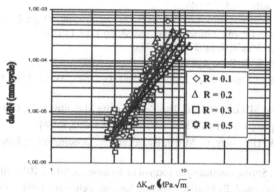

Figure 4. Evolution of FCGR for Aluminum alloy 2024 T351 as function of ΔK_{eff}

CONCLUSIONS

Main conclusions obtained in this study are summarized as follows:

- The fatigue life was affected by stress ratio. An increasing of stress ratio increase the fatigue life in the case when maximum applied is approximately constant.

- At low stress intensity factor we notice a shift of FCG curves to law threshold stress intensity, this situation shown an increasing of FCGR. At high stress intensity factor, we show a decreasing of FCGR (da/dN) when load ratio increase. This case shows the effect of microstructure in T-S orientation. Analysis of fracture surfaces for different load ratio has a future work that must demonstrate the effect of microstructure on FCGR.

- In order to have a good calibration and obtain an intrinsic curve of aluminum alloy, coefficients of Elber model must be adjusted.

ACKNOWLEDGMENTS

The authors gratefully acknowledge the technical support from Center of Material: Pierre-Marie Fourt, Paris and then wish to express their gratitude to Professor André Pineau, Dr. Benoit Tanguy for their contributions to the achievements of fatigue tests and Mme Anne Laurent for fractographic examination on SEM.

REFERENCES

1. P.C. Paris, M.P. Gomez, W.P. Anderson, The Trend Eng, **13**, pp 9-14, (1961).
2. R. Jones, L. Molent, S. Pitt, E. Siores, 2006. "Recent developments in fatigue crack growth", In: Gdoutos EE, editor. Proceedings of the 16th European conference on fracture, failure analysis of nano and engineering materials and structures, July 3-7, Alexandroupolis, Greece.
3. S. Dinda, D. Kujawski, Eng. Fract. Mech., **71**, pp 779-790, (2004).
4. G. Glinka, D. Kujawski, T. Tsakalakos, M. Croft, R. Holtz, K. Sadananda, 2004. "Analysis of fatigue crack growth using two driving force parameters", In: Proceedings of the international conference on fatigue damage of structural materials V, September 19-24, Hyannis, Massachusetts, USA, (2004).
5. R.G. Forman, V.E. Kearney, R.M. Engle, J. of Basic Engineering, **89**, pp.459-464, (1967).
6. E.K. Walker, ASTM **STP 462**. Philadelphia: ASTM, pp.1-14, (1970).
7. W. Elber, Eng. Fract. Mech., **2**, pp 37-45, (1970).
8. V. Sinha, C. Mercer, W.O. Soboyejo, Mat. Scie. Engng **A287**, pp 30-42, (2000).
9. S. Hariprasad, S. M. L. Sastry, K. L. Jerina and R. J. Lederich, Metal. Mat. Trans. A. **25(5)**, (1994).
10. AL. TH. Kermanidis, SP.G. Pantelakis, Fat. Fract. Engng Mat. Struct. **24**, 679-710, (2001).
11. M. Katcher, M. Kaplan, ASTM **STP 559**, ASTM, pp. 264–292, (1974).
12. R.J. Stofanak, , R.W. Hertzberg, G. Miller, R. Jaccard, K. Donald, Engng Fract. Mech. **17**, pp 527-539, (1983)
13. F.J. McMaster, D.J. Smith, International Journal of Fatigue **23**, S93–S101, (2001)
14. C.S. Kusko, J.N. Dupont, A.R. Marder, Welding Journal, **February 2004**, 59S-64S, (2004)
15. E.U. Lee, G. Glinka, A.K. Vasudevan, N. Iyyer, N.D. Phan, International Journal of Fatigue **31**, pp 1858-1864, (2009).
16. Y. Murakami. Stress intensity factors handbook. Pergamon Press, Oxford; 1: 9-17, (1987).
17. M. Benachour, M. Benguediab, A. Hadjoui, F. Hadjoui, N. Benachour, Computational Materials Science **44**, pp 489-495, (2008).

Mater. Res. Soc. Symp. Proc. Vol. 1276 © 2010 Materials Research Society

Influence of Tempering Temperature in Wear of AISI T15 HSS Tools Produced by HIP and Liquid Phase Vacuum Sintering

Emmanuel P. R. Lima[1], Maurício D. M. das Neves[2], Sérgio Delijaicov[2], Francisco A. Filho[2]

[1] UnB – FGA – Universidade de Brasília – Faculdade do Gama, DF/Brazil CEP: 72405-610.
[2] Centro Univer. da FEI – UniFEI, Assunção, S. Bernardo do Campo/SP/Brazil CEP: 09850-901.

ABSTRACT

This work aims to investigate the influence of tempering temperature in the wear resistance of AISI T15 HSS tools produced by two different sintering processes – hot isostatic pressing (HIP) and liquid phase vacuum sintering. All materials are submitted to annealing at 870°C, quenching at 1210°C and triple tempering at 540, 550 and 560 °C. Density measurements, hardness and bend strength (transversal rupture strength – TRS) tests are accomplished. To identify the present phases and to evaluate the obtained microstructures, analysis in optical microscopy, SEM and EDX are done. Interchangeable inserts are manufactured by electrical discharge machining process. Frontal machining without coolant of normalized AISI 1045 steel plates is employed. The cutting forces are monitored via a transducer basically constituted of an instrumented table with four load cells mounted with "Strain Gages" sensors capable to measure the cutting efforts. The tools wear is analyzed and used to estimate the performance of two different HSS tools. For both investigated materials, the tools tempered at 540 °C show the lowest wearing.

INTRODUCTION

The wear resistance is directly related to the cutting efficiency of high speed steel. This property depends on the tool hardness, alloy composition and carbides type. In general, the sintered high speed steels have a higher hardness than the conventional materials, besides a higher content of carbon and vanadium provides an increase of wear resistance mainly when thermically treated. The AISI T15 high speed steel is thermically treated similarly as the steels obtained by casting, molding and forming. This means quenching followed by multiple temperings to obtain the highest possible hardness. Multiple cycles of tempering reduce the amount of retained austenite [1] and precipitation of secondary carbides (M_2C) is promoted. They are responsible for the secondary hardness peak of high speed steels [2]. In general, sintered high speed steels tend to develop more predictable properties due to its more refined and uniform microstructure. During heat treatment independently of the processing sequences, the following processes will take place: carbides dissolution, proeutectoid carbides precipitation, transformation of the austenite in martensite and carbides precipitation in martensite.

This work aims to investigate the influence of the tempering temperatures on the flank wear of AISI T15 HSS tools produced by two different sintering processes, after frontal machining with normalized AISI 1045 steel plates in a dry condition,.

MATERIALS AND METHODS

The AISI T15 high speed steel powder is supplied by Coldstream Inc., while a commercial hot isostatically pressed (HIP) high speed steel is acquired, in billet's form, from

Eramet Latin America Inc. The chemical compositions of the two high speed steels are shown in table I.

Table I. Chemical composition of AISI T15 high speed steel [Wt.%]

Process Elements	C	W	Co	V	Cr	Mo	Si	Fe
Vacuum sintering	1.59	12.08	4.95	4.91	4.05	0.82	0.28	Bal.
HIP	1.56	12.00	4.97	4.93	3.91	0.43	0.50	Bal.

Samples of high speed steel powder are uniaxially cold pressed in metallic mold at a pressure of approximately 700 MPa and vacuum sintered at a temperature of 1275°C (± 3°C) for one hour in the presence of a liquid phase. Subsequently, samples of the two materials (commercial and vacuum sintered) are subject to heat treatments: annealing at 870°C, austenitizing at 1210°C (air-quenched) and triple tempering at 540, 550 and 560°C. All heat treatments are done in a salt bath. Densities are calculated before and after the sintering process.

After heat treating, the materials are metallographically prepared for microstructural analysis by SEM and EDX. Rockwell C hardness and flexion test in tree points (bend strength – TRS) are used to investigate the effectiveness of heat treatments and the materials toughness, respectively. The average grain size, volume fraction and diameter of the produced particles are determined by examining micrographs using a digital analysis method (Quantikov).

In a second stage, interchangeable inserts have been manufactured by electrical discharge machining according to ISO 1832-1977 standard specifications [3] with the purpose to evaluate its performance on the basis of tool wear. The frontal machining without coolant of annealed and normalized AISI 1045 steel plates is employed. The cutting forces are monitored via transducer basically constituted of an instrumented table with four load cells mounted with "Strain Gages" sensors capable to measure the cutting efforts in three orthogonal directions [4]. Cutting forces, surface roughness (R_a) and tool wear (VB) are used to estimate the performance of two different HSS tools. The tool life criterion is the flank wear (VB) being 0.90 mm. The machining parameters are the same for both materials.

RESULTS AND DISCUSSION

The obtained density values for the pressed samples before and after sintering are: 6.19 ± 0.08 g/cm^3 and 8.07 ± 0.05 g/cm^3, that mean 75.21% and 98.06% of the HSS AISI T15 theoretical density, respectively. The commercial material (HIP) presents a density of 8.18 ± 0.02 g/cm^3 (99.39%). As expected, the density of the commercial samples shows better (higher) values due to the largest efficiency of this process in the elimination of porosity [5].

Figure 1 shows SEM images of vacuum sintered AISI T15 high speed steel after quenching from 1210 °C and triple tempering at (a) 540 °C, (b) 550 °C, and (c) 560 °C. It contains MC and M_6C type carbides in a tempered martensite matrix. In addition, the amount and size of the MC type carbides (in gray tones) is larger than that of the eutectic M_6C type carbides (in white). This can be justified by the low solubility of the MC [6] carbides during the austenitizing process and also due to the low diffusion rate during tempering treatment. In such a case, a relatively low temperature is used which can prevent dissolution of carbides and a subsequent re-precipitation stage.

Figure 1. SEM images of vacuum sintered HSS AISI T15 samples quenched at 1210 °C and triple tempered at (a) 540 °C, (b) 550 °C and (c) 560 °C.

Sample tempering at 550°C produces little significant differences regarding carbide morphology. However, a larger dispersion is found in the spatial distribution and size of the carbides particles (figure 1b). Figure 1b also shows that the material microstructure is heterogeneous. This behavior can be explained by the differences in pressure generated by the large pressing section area (59 mm in diameter) in use. This can generate volumes with different densities.

Samples tempering at 560°C gives rise to large size dispersion of the carbides, especially for MC type carbides (see Figure 1c). The particle spatial distribution is also more uneven . There are areas with a considerably higher number density others where larger carbide particles are found. This result suggests that for this tempering temperature (560 °C), dissolution and coalescing (coarsening process) of the smaller carbides (M_6C) take place simultaneously with the growth and precipitation of the MC and M_2C type carbides [7]. Regarding grain size, practically there is not alteration.

Figure 2. SEM images of HIP sintered AISI T15 HSS samples quenched at 1210 °C and triple tempered at (a) 540 °C, (b) 550 °C and (c) 560 °C.

The commercial AISI T15 high speed steel present considerably smaller grain and carbides particles sizes (Figure 2a) as compared to vacuum sintering high speed steel. Figure 2a shows a uniform carbide size distribution (MC and M_6C) in the matrix, as well as, a higher carbide number density for the M_6C type (in bright tone). The corresponding morphology is also rather uniform. In spite of the high measured density, there is a noticeable residual porosity (see Figure 2a).

As for the triple tempered material (550 °C), Fig. 2b shows that there are few and non significant changes in the morphology and distribution of the carbide particles. There is only a discreet increase in grain size when compared to the material treated at 540 °C.

There is a small increase of the number density of M_6C (in white colour) type carbides for HIP sintering AISI T15 high speed steel tempered at 560°C (figure 2c). In spite of the high uniformity of the commercial material, the small difference in grain sizes as a function of the tempering temperatures can be attributed to microstructural heterogeneity. This can be due to the relatively large diameter of the compacted (diameter of 90 mm) material, producing a variation of density between the center and the border of the billet [5].

Table II. Values of austenitic grain size, carbides mean diameter, Rockwell C hardness and TRS for the AISI T15 HSS heat treated at 1210 °C (air-quenched).

Material Tempering	Grain size [μm]	Φ Carbides [μm]	Hardness [HRc]	TRS [MPa]
Vacuum sintering 540°C	19.75 ± 4.89	1.13 ± 0.90	65.36 ± 0.32	1409 ± 148
Vacuum sintering 550°C	20.59 ± 6.49	1.05 ± 0.71	64.50 ± 0.27	1445 ± 101
Vacuum sintering 560°C	20.94 ± 6.82	1.08 ± 0.74	63.91 ± 0.23	1579 ± 180
Commercial (HIP) 540°C	09.25 ± 2.12	0.68 ± 0.28	67.41 ± 0.25	2050 ± 82
Commercial (HIP) 550°C	10.70 ± 1.91	0.75 ± 0.21	66.86 ± 0.41	2180 ± 79
Commercial (HIP) 560°C	10.84 ± 2.00	0.77 ± 0.23	65.90 ± 0.35	2312 ± 92

Table II shows a slight decrease of hardness as the tempering temperature increases in the two differently sintered HSS tools under evaluation. An inverse behavior is found for the bend strength (TRS). This is similar to results in a previous work by Lima *et al.* [8], suggesting that there is an interval in which highest hardness is reached (secondary peak of hardness) at a temperature below 550°C. Thus, the increase of the tempering temperature only contributes to relief internal tensions developed during austenitizing (air-quenched) and produces an increase of the material ductility. It is also shown that the hot isostatic pressed high speed steel presents the highest hardness and TRS values. This behavior can be justified by the smallest grain size and the highest microstructural uniformity (morphology and distribution) in this material.

As it is known, the increase of the cutting forces is mainly related to progress in tool wear during machining. Figure 3 shows that increasing the milled length produces very clear differences in the cutting forces among the triple tempered tools at 540, 550 and 560°C. For both materials and with respect to the cutting forces, the best results are found at a tempering temperature of 540 °C.

A similar behavior is for the flank wear – VB (Figure 4). The milled lengths produced by the vacuum sintered and triple tempered tools produce the following milled lengths: 1800, 1600 and 1400 mm, respectively (see Fig. 4a). HIP sintering AISI T15 HSS tools present the best performance when subject to heat treatment (triple tempering at 540 °C), reaching a milling length of 2000 mm until to point that the tool is considered worn. The tools treated at 550 and 560 °C mill respectively 1800 and 1600 mm (Fig. 4b).

As for the surface finish of the milled part, Figure 5 shows hat the roughness varies significantly as a function of the tempering temperature of the tools but only for distances around 1400 mm i.e. the end of tool life. In this manner and for the used conditions in this work, the choice of the tempering temperature is not decisive.

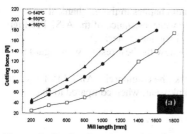 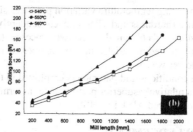

Figure 3. Cutting forces as a function of machining length monitored during milling operation of AISI T15 HSS tools: (a) vacuum sintering, (b) commercial.

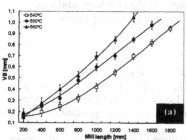 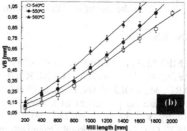

Figure 4. Curves of flank wear (VB) versus machining length monitored during milling operation of AISI T15 HSS tools: (a) vacuum sintering, (b) commercial.

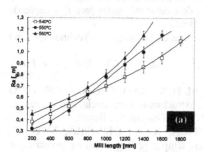 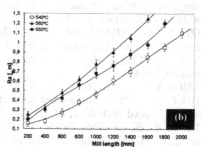

Figure 5. Curves of surface roughness (R_a) versus machining length monitored during milling operation of AISI T15 HSS tools: (a) vacuum sintering, (b) commercial.

CONCLUSION

1. No occurrence of chipping or catastrophic failure of tools is observed in all machining trials under the investigated conditions.
2. The higher the tool wear is, the higher the involved cutting forces are;

65

3. The tempering temperature variation produces significant changes in the mechanical properties (hardness and TRS) and, consequently, in the wear resistance of the AISI T15 high speed steel tools obtained by the two processes (HIP and vacuum sintering);
4. For both studied materials, the tools tempered at 540°C show the best results with regard to wear (lower wear);
5. For the conditions and parameters employed in this work, the commercial AISI T15 HSS tool shows a relatively superior performance, in terms of flank wear, when compared to the vacuum sintered AISI T15 HSS tool.

ACKNOWLEDGEMENTS

The authors thank DPP-UnB for the financial support for participation in this event, to IPT for the pressing of the materials, the company Hurth Infer for the heat treatments and, finally, to UniFEI for the use of his laboratory for measurement of the cutting forces.

REFERENCES

[1] ASM METALS HANDBOOK. (1991). "Heat treating", Vol. 4, pp. 734-760.

[2] BREWIN, P. R. TOLOUI, B., NURTHEN, P. D., FELLGET, J. A., WOOD, J. V., IGHARO, M., COLEMAN, D. S. and SHAIKH, Q. (1989). "Effect of process variables and microstructure on properties of sintered high speed steel for wear applications". Powder Metallurgy, Vol. 32, N° 4, pp. 285-290.

[3] ISO 1832 (1977). International Organization for Standardization – Indexable (throwaway) inserts for cutting tools.

[4] ROSSI, G. C., DELIJAICOV, S., BORDINASSI, E. C., BATALHA, G. F. (2007). "Estudo das forças de corte no processo de corte de bordas de chapas utilizadas para fabricação de tubos de aço com costura". COBEF, São Pedro-SP, Brazil

[5] KOIZUMI, M.; NISHIHARA, M. (1991). "Isostatic Pressing – Technology and Aplications". Elsevier, London and New York.

[6] HOYLE, G. (1988). "High speed steel". Butterworth & Co, The University Press, Cambridge, London.

[7] NOGUEIRA, R. A., RIBEIRO, O. C. S., NEVES, M. D. M., SALGADO, L., AMBROZIO FILHO, F. (2003). "Effect of heat treatment on microstructure of commercial and vacuum of sintered high speed steels AISI M2 and T15", IV PTECH, Guarujá-SP, Brazil.

[8] LIMA, E. P. R., NEVES, M. D. M., NOGUEIRA, R. A., OLIVEIRA, L. G. C., AMBROZIO FILHO, F. (2007). "Effect of different tempering stages and temperatures on microstructure, tenacity and hardness of vacuum sintered HSS AISI T15", VI PTECH, Búzios-RJ, Brazil.

Mater. Res. Soc. Symp. Proc. Vol. 1276 © 2010 Materials Research Society

Effect of Hot Band Annealing on the Microstructure and Mechanical Properties of Low Carbon Electrical Steels

E. Gutiérrez-Castañeda[a], A. Salinas-Rodríguez [b].

Centro de Investigación y de Estudios Avanzados del IPN, P.O. Box 663, Saltillo, Coahuila, México, 25000. [a]emmanuel.gutierrez@yahoo.com.mx, [b]armando.salinas@cinvestav.edu.mx

Keywords: Electrical steels, hot band annealing, microstructure, mechanical properties.

ABSTRACT

Magnetic properties of grain non-oriented low-C electrical steels are improved when the hot rolled strip is annealed (HBA) prior to cold rolling and final annealing treatments. This improvement results from development of a {100}<uvw> texture in the large grained ferrite microstructure produced during final annealing. HBA at 800-850 °C results in rapid decarburization and elimination of carbide particles which have caused concerns about the suitability of the mechanical properties in the final product. In this work, samples taken from a hot rolled electrical steel coil are subjected to HBA during 150 minutes at 850 °C, cold rolled and finally annealed three minutes at temperatures between 700 and 1000 °C. The resulting tensile properties are compared with those of samples subjected to a similar processing route but without the HBA treatment and samples of industrially semi-processed grain non-oriented electrical steel decarburized 16 hours at 750 °C. It is shown that the yield strength of samples with and without HBA depends on the final grain size according to the Hall-Petch relationship; the final grain size depends strongly on annealing temperature. However, the HBA treatment causes the strength to decrease by a factor of about 2.5 and the ductility to increase by a factor of about 1.5. It is observed that the microstructure and tensile properties of the semi-processed electrical steel subjected to a final decarburization annealing are identical to those observed in material subjected to HBA in the present work. These results indicate that the HBA treatment not only improves the magnetic properties but also leads to a significant reduction of production time for grain non-oriented electrical steels.

INTRODUCTION

Grain non-oriented (GNO) electrical steel occupies the first place among magnetic materials in terms of both total tonnage and market value [1]. This material is used for the manufacture of cores of electrical machinery such as motors, generators, transformers, etc. For these applications, low energy losses and high permeability are required [2]. Along with good magnetic quality, the ability of the steel to be shaped by stamping is also increasingly important. This property to a large extent determines the service life of the expensive dies used in this operation, and increasing the life of the dies is an efficient way of reducing the cost of electrical equipment. Hence, mechanical properties of these materials are as important as magnetic properties [3, 4]. It is well known that properties of electrical steels are influenced strongly by the thermo-mechanical history [5, 6]. Conventionally, laminations of GNO electrical steels are processed from hot-rolled strip followed by cold rolling and continuous or batch annealing.

Finally, the strip is temper rolled and decarburized [7]. In this work, the microstructure of hot-rolled bands is modified by annealing (HBA) prior to cold rolling. Finally, the cold rolled samples are annealed for a short time. The microstructure, yield strength and ductility are compared to those of samples without this heat treatment and those obtained in production scale of fully processed electrical steels. It is observed that HBA not only improves the magnetic quality but also the mechanical properties of these materials. Moreover, there is a significantly reduction on the processing time caused by hot band annealing.

EXPERIMENTAL

Hot rolled bands, 2.3 mm in thickness, of commercial GNO electrical steel are obtained from a local steelmaker. Table I lists the chemical composition of the hot-rolled electrical steel bands. Rectangular samples, 10 cm wide x 15 cm long, are cut from the hot band. The samples are then heated at rate of 15 °C/min and annealed in still air at temperatures between 700 and 1050 °C in a muffle-type furnace. Annealing times are 30, 60, 120 and 150 min and cooling is performed in air. The as-received hot-rolled band and the annealed hot-rolled band samples are cold-rolled to a final thickness of 0.57 mm using a laboratory two-high rolling mill. Then, cold-rolled steel sheets, 3 cm wide x 30 cm long, are cut and annealed during 3 min at temperatures between 700 and 1000 °C. The microstructures of longitudinal sections of the resulting samples are characterized using optical and scanning electron microscopy. The grain size is measured using the Image Pro-plus program by the intercept method based on ASTM Standard E-112. The transformation temperatures are determined by dilatometry. Samples are heated at 10, 20, 30, 40 and 50 °C/min from room temperature up to 1200 °C. Soaking time is 10 min and an argon atmosphere is used through the whole set of experiments. Thus, the critical changes on the (de_L/dT) first derivative curve are obtained and then plotted vs heating rates. Finally, the mechacnical properties of samples resulting from final annealing are determined based on ASTM Standar E-8.

Table I. Chemical composition of the experimental steel [wt %]

C	Si	Al	S	Mn	P	Cu	Cr	Ni	N_2
0.05	0.57	0.21	0.004	0.32	0.042	0.029	0.019	0.029	0.0056

RESULTS AND DISCUSSION

Microstructure

Figs. 1a, 1b and 1c show the microstructural evolution of samples without HBA. Also, the microstructures of samples subjected to HBA at 850 °C during 150 min are shown (Figs. 1d, 1e and 1f). As can be seen, the grain structure is changed by HBA. The results of dilatometry experiments show that the Ac_1 and Ac_3 transformation temperatures increase as a function of the heating rate (Fig. 2a). It is observed that the transformation temperatures for a heating rate of 15 °C/min (the same used during HBA), are about 740 and 950 °C, respectively. Thus, annealing of hot-rolled bands at temperatures within the intercritical region, say 850 °C, causes rapid ferrite grain growth from the surface to the mid-plane of the sample resulting in abnormal ferrite columnar grains free of second-phase particles (Fig. 1d). The development of these columnar grains in non-oriented electrical steels has been reported by other researchers [8]. They found that columnar microstructures in the steels are formed only if the decarburisation annealing is performed in the α + γ region and a sharp interface exists between the decarburised ferrite and two-phase area. It is suggested that formation of the columnar microstructure would be expected when the velocity of the transformation front is higher than the nucleation speed of the phase transformation. Nevertheless, the decarburisation process in the two-phase region can be an efficient controlling mechanism for obtaining the columnar ferritic microstructure in the solid state [8]. Although the mechanism of these columnar grains is not presented in this work, it seems that fast decarburization at $Ac_1 < T < Ac_3$ favored this type of columnar growth (Fig. 1d).

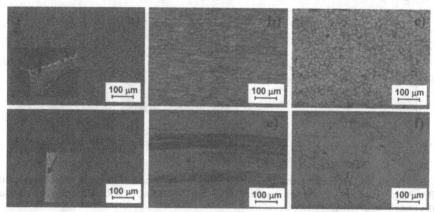

Figure 1. Microstructures of samples without (b and c) and with HBA at 850 °C during 150 min (d, e and f). (a) As received hot rolled band, (d) Sample subjected to HBA=850°C, 150 min, (b and d) Cold-rolled samples and (c and f) Samples annealed at 1000°C, 3 min.

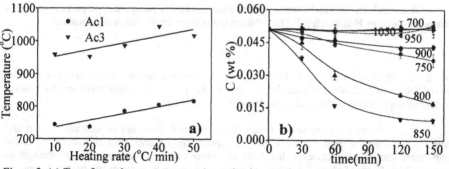

Figure 2. (a) Transformation temperatures determined by dilatometry and (b) evolution of carbon during hot band annealing treatment.

Fig. 2b shows that annealing of hot-rolled bands at 850°C causes decarburization of the strip in about 60 minutes. Cold-rolling of both the as-received hot-rolled band and the annealed hot band produces microstructures of severely deformed grains elongated parallel to the rolling direction. However, the length and thickness of the deformed grains depend on the size of the original grains produced prior to cold-rolling by hot band annealing (Figs. 1b and 1e). After final annealing, the samples subjected to HBA result in microstructures with larger grain size than those without this heat treatment, compare Figs. 1c and 1f. This is attributed to the lower amount of second-phase particles caused by decarburization during HBA, which favors recrystallization and grain growth during final annealing. Fig. 3 shows the effect of temperature of final annealing on the grain size. As can be seen, the grain size increases with increments in the annealing temperature. Nevertheless, the rate at which grain growth occurs is much faster in samples subjected to HBA.

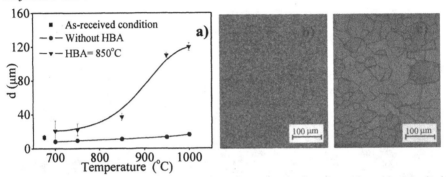

Figure 3. (a) Effect of temperature of final annealing on grain size of samples without and with HBA. Microstructures of samples resulting from final annealing at 850 °C, 3min: (b) without HBA and (c) with HBA.

Mechanical properties

A general relationship between yield strength (and other mechanical properties) and grain size is proposed by Hall-Petch [6]. This relationship is given by the following equation.

$$\sigma_o = \sigma_i + kD^{-1/2} \qquad [1]$$

where σ_o = the yield stress; σ_i = the friction stress, representing the overall resistance of the crystal lattice to dislocation movement; k = the locking parameter, which measures the relative hardening contribution of the grain boundaries; D = grain size.

Fig. 4 shows a comparison between the mechanical properties of samples without and with HBA. As can be seen, the yield strength depends on the final grain size according to the Hall-Petch relationship. However, hot band annealing treatment causes the yield strength to decrease by a factor of about 2.5 and the ductility to increase by a factor of about 1.5, Figs. 4a and 4b, respectively. The friction stress, which is a measure of the stress needed to move unlocked dislocations on the slip plane, is lower for samples subjected to hot band annealing. Therefore, when the amount of grain boundaries is lower (larger grain size), the amount of obstacles to dislocation slip and consequently the friction stress needed to move dislocations are lower. On the other hand, annealing within the ($\alpha + \gamma$) two phase field does not appear to have a significant affect on the ductility of the present steel. In contrast, annealing at 950 and 1000 °C does have a profound effect on ductility. As shown in Fig. 4b, the ductility first increases with increments in the annealing temperature up to 850 °C and then decreases at higher temperatures. The decrease in ductility can be associated to the increase in the amount of secondary microconstituents and to the increment of the inclusions size observed in samples annealed at $T \geq Ac_3$.

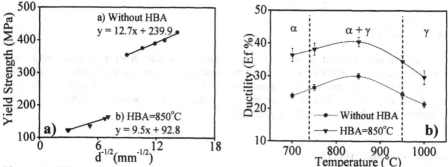

Figure 4. Effect of final annealing temperature on (a) the Hall-Petch relationship and (b) % elongation to fracture of samples without and with hot band annealing.

Comparison with the microstructure and mechanical properties produced industrially by the typical processing route

Annealing of hot rolled bands at temperatures within the intercritical region causes rapid decarburization, which helps recrystallization and grain growth during final annealing. As a result, large microstructures very similar to those observed in fully processed electrical steels are obtained. Fig. 5 shows a comparison between the microstructure of an industrially fully processed electrical steel fabricated by the typical processing route (hot rolling, cold rolling, intermediate annealing, temper rolling and decarburizing annealing at 750 °C during 16 hrs) with the microestructure of samples resulting from the alternative processing route (HBA at 850 °C during 150 min, cold rolling and final annealing at 1000 °C during 3 min), Figs. 5a and 5b, respectively. As can be seen, the ferrite grains (average grain size: 113 μm Vs 120 μm) are very similar. As a result, the yield strength (125 MPa vs 122 MPa) and the elongation to fracture (30% Vs 27%) are very similar. Therefore, mechanical properties of GNO electrical steels depend strongly on the microstructural features of steel. The amount of both carbon in solution and second phase particles as well as the final grain and inclusion size, all contribute to adjust the final mechanical properties of these materials.

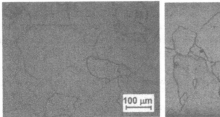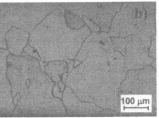

Figure 5. Comparison between the microstructures of GNO low C electrical steels fabricated by the (a) typical processing route and (b) alternative processing route.

CONCLUSIONS

The results obtained during this investigation show that, from a microstructural and mechanical properties point of view, hot band annealing technology represents an attractive alternative processing route for the manufacture of fully processed electrical steels strips. The productivity of these steels can be increased significantly considering the reduction on the production time by hot band annealing process.

ACKNOWLEDGEMENTS

The authors want to thank CONACyT for supporting this work under grant 166596. Also, the collaboration of Rogelio Deaquino Lara, Martha Rivas and Felipe Márquez in this investigation is duly recognized.

REFERENCES

[1] A.J Moses: Past, Present and Future Developments, IEE Proceeding Vol. 137, p. 233 (1990).
[2] A., Yoshio, S. Okamura, History and Recent Development of Non-Oriented Electrical Steel, Technical report, 39, p.13-20 (1998).
[3] Philip Beckley, Electrical Steels for Rotating Machines, Mechanical Properties, London, p. 169 (2002).
[4] L. Prisekina, L. Kazadzhan, Y. Larin, N. Krutskikh, Effect of Processing Parameters on the Mechanical Characteristics of Isotropic Electrical Steel, J. Metallurgist, Vol. 32, No. 4, p. 137-139 (1988).
[5] Kim Verbeken, Edgar Gomes, Juergen Schneider and Yvan Houbaert, J. Solid and State Phenomena , Vol. 160, p. 189-194 (2010).
[6] George Ellwood Dieter, Mechanical Metallurgy, Hall-Petch Relation and Annealing of Cold Worked Metal, Massachusetts, p. 189, 233 (1986).
[7] J.W. Lee, P. R. Howell, Microstructural Development in Non-oriented Lamination Steels, J. Mater Sci, 22, p. 3631-3642 (1987).
[8] Mykola Dzubinsky, Yuriy Sidor, Frantisek Kovac, Kinetics of Columnar Abnormal Grain Growth in Low-Si Non-oriented Electrical Steel, J. Mater. Sci. Eng., 385, p. 449-454 (2003).

Mater. Res. Soc. Symp. Proc. Vol. 1276 © 2010 Materials Research Society

Microstructural and Mechanical Characterization of a TRIP-800 Steel Welded By Laser-CO_2 Process

G.Y. Perez-Medina[1], P. Zambrano[2], H.F. López [1], F.A. Reyes-Valdés[1], V. H. López-Cortés[1],
[1] Corporación Mexicana de Investigación en Materiales. Calle Ciencia y Tecnología #790, Fracc. Saltillo 400, Saltillo, Coah. México 25290.

[2] Universidad Autónoma de Nuevo León. Facultad de Ingeniería Mecánica y Eléctrica. Av. Pedro de Alba S/N. Col Ciudad Universitaria. San Nicolás de los Garza Nuevo León.

ABSTRACT

This paper presents results on the impact of Laser CO_2 process variables on the weldability, phase transformations and tensile properties of a TRIP800 Steel. The microstructure of this steel is comprised of ferrite, bainite and retained austenite phases. This is obtained by controlled cooling from the intercritical annealing temperature to the isothermal bainitic holding temperature. These steels have been increasingly used in the last 10 years in the automotive industry and for these materials to be used effectively; the influence of material and the CO_2 laser welding process condition must be clearly understood. Hence, in this work the effect of the welding process on the resultant microstructures and on the exhibited mechanical properties is investigated. It is found that the tensile strength of welded specimens falls below 800 MPa and that the elongation becomes 15 % or lower. In turn, this clearly indicates that the implemented laser welding process leads to a reduction in the TRIP800 steel toughness.

KEYWORDS: AHSS, TRIP Steel, GMAW, Laser, welding

INTRODUCTION

The demand for high efficiency vehicles has lead to the development of advanced steels with unique mechanical properties. Among these steels, the advanced high strength steels (AHSS) have been developed for the manufacture of low weight car bodies using thin sheet gauge of less than 1 mm [1]. The AHSS for applications in the automotive sector include transformation induced plasticity (TRIP) steels [1]. TRIP steels possess a microstructure consisting of a ferrite (F) and bainite (B), including retained austenite (RA) phases. The enhanced plastic behavior of TRIP steels is attributed to the strain induced transformation of the retained austenite into martensite during deformation. These steels exhibit relatively high work hardening rates and remarkable formability. In addition, the level of applied plastic strain needed to induce the austenite to martensite transformation is strongly dependent on the carbon content. At low carbon levels, the retained austenite transforms almost immediately once the material reaches the yield strength. At elevated carbon contents, the retained austenite becomes increasingly stable and can only transform at elevated levels of plastic straining such as the ones found during a sudden crash event [2]. The formation of unwanted martensite is among the main concerns related to welding [3, 4]. It has been found that in low heat input welding processes such as resistance spot welding (RSW), high carbon martensites can form in the weld that lead to embrittlement [3, 4].

Accordingly, in this work the effects of welding on the resultant microstructures and on the mechanical integrity of a TRIP steel are investigated using low heat input processes. For this purpose, a Laser CO_2 welding process is employed for welding a TRIP thin sheet steel currently used in the automotive sector.

EXPERIMENT

The chemical composition of the TRIP steel is given in Table I. Table II shows the mechanical properties of the TRIP steel in the form of 1.6 mm thick sheet. Tensile bars have been cut from the steel sheet, each having a size 244 x 70 x 1.6 mm and are welded by a Laser CO_2 process. The welding parameters in the Laser CO_2 process are given in Table III. The used equipment is a CO_2-LBW-unit, EL.EN-RTM of 6 kW with 6 degrees of freedom. Optical and scanning electron microscopy (SEM) and X Ray Diffraction (XRD) are employed in characterizing the exhibited microstructures and fracture modes of the as-welded TRIP steels. The hardness of the various welding regions is determined by Vickers micro-hardness profile determinations. Finally, tensile strength and ductility of the welded strips are determined by using a universal testing machine.

Table I. - Chemical composition of the AHSS TRIP 800.

Wt%	C	Mn	Si	P	Al	Cu	Cr	Ni	Mo	Sn
TRIP800	0.232	1.653	1.55	0.010	0.041	0.033	0.033	0.036	0.018	0.006

Table II. Mechanical properties of the AHSS TRIP 800

Base Metal	Yield Strength [MPa]	Ultimate Tensile Strength [MPa]	Elongation [%]
TRIP800	450	800	28

Table III.- Welding Parameters using CO_2 laser processes in a TRIP800 steel.

Welding Process	Joint type	Power [kW]	Welding speed [mm min^{-1}]	Heat Input [J mm^{-1}]
Laser CO_2	Butt Joint	4.5	3699.76	72.972

RESULTS AND DISCUSSION

Figure 1a shows the microstructural features of the TRIP steel in the as received condition. Notice the mixture of ferrite and bainite phases and possibly some residual austenite. The resultant microstructures in the heat affected zone (HAZ) and fusion zone (FZ) of the welded TRIP800 steel are shown in Figure 1b and 1c respectively. From these micrographs it is evident that the microstructures in the steel consist of martensite (M), lower bainite (LB), ferrite (F) and retained austenite (RA) phases in the HAZ. The fusion zone (FZ) is formed by the same phases in different proportions.

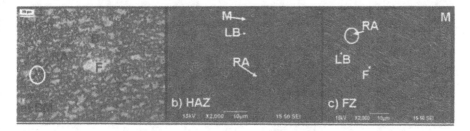

Fig. 1. Microstructures of the AHSS TRIP800 in (a) As-received material, (b) HAZ and (c) FZ.

The microhardness profiles for the various regions of the welded TRIP800 steel are shown in Figure 2a. In this figure the numbers represent locations of indentations taken at spacing intervals of 1 mm. Microhardness measurements indicate that welding promotes a significant increase in hardness in the welded regions. In particular, it is found that the metal in the regions adjacent to the HAZ, exhibits a significant increase in hardness (points 4 and 11 in Fig. 2a). Apparently, due to the relatively high cooling rates, possible strain induced martensite (SIM) and the development of residual stresses in the as received (BM) region adjacent to the HAZ as there are no clear phase transitions identified in this region. There are microhardness increments from 275 HV in the as received condition up to 500 HV in the HAZ and over 600 HV in the FZ. In particular, the microhardness profiles acquire a "top hat" morphology [7] with a maximum hardness of 600 HV in the parting line. These microhardness profiles are attributed to the development of an athermal martensite phase [7] which is no longer a function of cooling rate.

The volume fraction of retained austenite in the BM, FZ and HAZ is determined by image analysis (Figures 2b-d) and X-ray diffraction (XRD) (Figure 3) by using Cu Kα radiation. An estimation of the Retained austenite is calculated using the following equation [8]:

$$V_\alpha = \frac{1.4 I_\gamma}{I_\alpha + 1.4 I_\gamma} \tag{1}$$

Where I_γ is the average intensity of the peaks of austenite and I_α is the highest intensity of the peak of ferrite. The XRD results and the applied parameters are given in table IV. It can be seen that the percentage of RA in the HAZ and FZ reduce to 11.29% in comparison with the 12.54 % in the BM. This suggests the transformation of RA to martensite. This can be further corroborated by quantitative image analysis as RA decreases 13 % in HAZ and 7% in FZ.

The low heat input in the Laser CO_2 process allows a cooling rate around 417.81° K/s [9]. From the current results, it is clear that laser welding gives rise to relatively fast cooling rates. Critical cooling rates for the transformation of austenite to martensite can be determined from continuous cooling transformation (CCT) diagrams. Unfortunately there are no reports on the CCT curves for the TRIP800 steel. Nevertheless, Li [5] and Badeshia [6], have proposed thermodynamically and kinetically based models for predicting CCT diagrams in a wide range of steels [6]. From these estimations, it is found that in the present steel the critical cooling rates for the formation of martensite are between 45 and 90 °C/s.

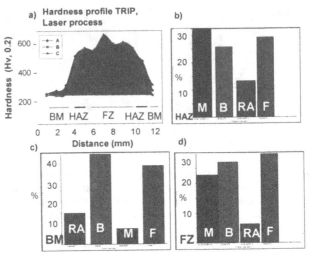

Fig. 2. (a) Hardness Profiles. Phases percentage in (b) HAZ, (c) As received condition and (d) FZ.

Table IV.- XDR results and parameters applied for CO_2 laser processes in a TRIP800 steel.

Sample code	Intensity (I)					RA (%)
	α-Iron	γ-Iron				
Base Metal	(100)-1000	(220)-330	(311)-40	(222)-20	(400)-20	12.54
FZ and HAZ	(101)-660	(200)-70	(220)-50	-	-	11.29

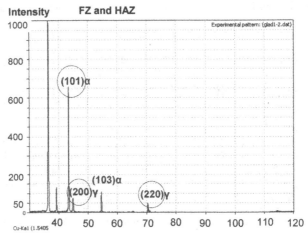

Fig. 3 shows the measured XRD data for FZ and HAZ.

The ultimate tensile strength (UTS) and ductility of the welded TRIP800 steels are shown in table V. The maximum UTS values are below 800 MPa and the elongation drops down to 15%. In turn, this clearly indicates that laser welding of TRIP800 steels leads to a reduction in the steel toughness. Confirmation for the loss of toughness is found by looking that the fracture occurrs in the BM regions adjacent to the HAZ. Also, the fracture appearance is brittle cleavage possibly chevron markings (see Figs. 4a-b). The fracture surfaces are relatively flat and there is a lack of appreciable cavitation and thus insufficient ductility. Although, the BM region has a potentially ductile microstructure consisting of a ferritic matrix with bainite and retained austenite, it becomes susceptible to fracture as a result of (a) SIM driven by internal stressing and (b) the development of possible residual/internal stresses at this location as a result of the fast cooling rates of the process. As Fig. 2a shows, the BM region hardness values increase appreciably in the neighborhood of the HAZ. This suggests the development of internal stresses and/or the formation of martensite from any residual austenite. A comparison of the resultant microstructures in the BM region adjacent to the HAZ with the one away from the HAZ is given in Fig. 5. Notice from these figures that appreciable coarsening of the various phase constituents occurs in the BM adjacent to the HAZ. Accordingly, it is apparent that in the HAZ of the Laser CO_2 welded strips, tempering effects of the bainite/martensite phases coupled with phase coarsening and residual stresses including SIM promote brittleness. It is at the moment uncertain whether or not such an embrittlement can be avoided by a careful selection of the processing parameters during Laser welding and it requires further investigation.

Table V. Tensile test results of TRIP800 Welded by laser CO_2

No. Sample	% Elongation	Ultimate Strength Mpa.	Fracture Zone
1	15.00	748.43	HAZ
2	21.67	789.83	BM
3	15.00	767.37	HAZ

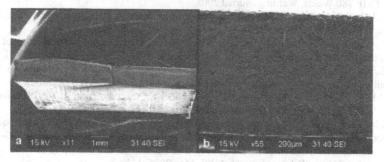

Fig. 4. (a) Overall view of the fracture surfaces; (b) brittle fracture appearance typical of cleavage with apparently river markings.

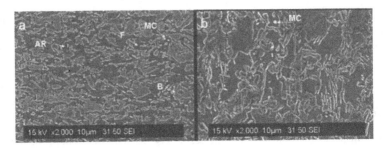

Fig. 5. Comparison of the resultant microstructures in the BM region of the Laser CO_2 process,
(a) BM region adjacent to the HAZ, (b) BM region away from the HAZ.

CONCLUSIONS

The weldability of a thin sheet of a TRIP 800 steel using Laser CO_2 process has been investigated by using microhardness measurements in combination with XRD and tensile testing. It is found that welding using Laser results in martensitic structures in the HAZ and a mixture of martensite, bainite, ferrite and RA phases in the FZ. The FZ including the HAZ are relatively hard compared with the BM. The measured mechanical properties indicate that the BM region adjacent to the HAZ undergoes brittle fracture. Apparently, tempering of the phases in this region results in poor ductility by promoting phase coarsening. The formability and the amount of RA decrease in HAZ and FZ. Based on XRD analysis, the volume fraction of RA and the percentage of elongation in HAZ were 11.29 and 15%.

REFERENCES
1. BY N. Kapustka. C. Conrardy. Effect of GMAW process and Material Conditions on DP 780 and TRIP 780 Welds, Welding Journal 2008.
2. I.D.Choi et al. (2002), Deformation behaviour of low carbon TRIP sheet steels at high strain rates. *ISIJ Int 2002*;42(12):1483–9
3. J. E., Gould, L. R. Lehman, S. Holmes, (1996). A design of experiments evaluation of factors affecting the RSW of high-strength steels. Proc. Sheet Metal Welding Conference VII, AWS
4. J. E., Gould, D. Workman, (1998), Fracture morphologies of RSW exhibiting hold time sensitivity behavior. *Proc. Sheet Metal Welding Conference VIII*, AWS Detroit S.
5. Li, M. V; Niebuhr, 1998. A computational model for the prediction of steel hardenability. Metallurgical and Materials Transactions 29B (6):661.
6. Bhadeshia, H. K. D. H; and Svensson, L-E 1993. Mathematical Modeling of Weld Phenomena, eds, H. Cerjack and K. E. Easterling, Institute of Metals, London, pp.109-180.
7. J.E. Gould, S.P. Khurana, T. Li; (2006), Predictions of microstructures when welding automotive AHSS; *Welding Journal*, AWS, May 2006, 111.
8. M. De Meyer, D. B.C.D. Cooman 41st MWSP Conference Proceedings, ISS, 1999, pp.483.
9. G.Y. Perez-Medina, F.A. Reyes-Valdés, H. F. Lopez, Structural Integrity of a Welded TRIP800 Steel Using Laser CO_2 and GMAW Processes; Rivista Italiana della Saldatura N-3 2010, pp. 333-338.

Mater. Res. Soc. Symp. Proc. Vol. 1276 © 2010 Materials Research Society

Mechanism of Grain Growth During Annealing of Si-Al Electrical Steel Strips Deformed in Tension

J. Salinas B.[1], A. Salinas R.[2]
Centro de Investigación y Estúdios Avanzados del IPN. P. O. Box 663, Saltillo Coahuila,
México 25000. E-mails: jorgesb231182@yahoo.com.mx[1], armando.salinas@cinvestav.edu.mx

ABSTRACT

The mechanism of recrystallization as a result of annealing during 600-7200 seconds at 700 °C of a Si-Al, low C electrical steel strip is investigated in samples deformed in tension. The evolution of grain size during annealing is evaluated by optical microscopy and electron backscatter diffraction in the scanning electron microscope. It is found that grain growth starts after an incubation time of 600 s with no apparent evidence of primary recrystallization. After that, the grain size-time relationship exhibits two different stages. Initially, the grain size increases linearly with time up to about 3600 s. During this time, some selected grains grow until they consume the deformed microstructure. In the second stage, the rate of growth decreases significantly and a final grain size of about 150 μm is reached after 7200 seconds of annealing. Grain orientation spread maps obtained from EBSD data of deformed and partially recrystallized samples during the stage of linear growth reveals that the growing grains exhibit lower misorientation and therefore smaller stored energy than the non-recrystallized matrix grains. Analysis of image quality maps reveal that the IQ values for {100}<uvw> orientations are higher than those observed for {111}<uvw> orientations thus suggesting that the {100}<uvw> orientations grow at the expense of {111}<uvw> orientations by a mechanism of strain-induced boundary migration.

INTRODUCTION

Non-oriented electrical steels are used in motors and transformer cores due their low energy losses [1]. Their magnetic properties, such as magnetization curves, permeability, coercive force and specific magnetic energy losses are all related to their microstructure [2, 3] which must be strictly controlled during fabrication. Electrical steel strips are processed like most cold rolled steels: hot rolling, cold rolling, box annealing and temper rolling [4].

In the case of semi-processed grain-non oriented electrical steel strips, the final processing stage at the steelmaking plant is a small deformation by rolling known as "skin-pass" or "temper rolling". The final magnetic properties are then developed during an annealing treatment conducted after punching the laminations at the motor/transformer fabrication plant. The temper rolling strain is of paramount importance in manufacturing this type of steel strips because the final grain size and, therefore, the resulting magnetic properties, depend on it. In the present study the mechanisms of recrystallization and grain growth during annealing of samples obtained from a low-C, Si-Al electrical steel strip deformed to small tensile strains is investigated using orientation imaging microscopy. Image quality and intragranular misorientation maps are analyzed to qualitatively estimate differences in energy stored for various orientations present in the deformed samples.

EXPERIMENTAL PROCEDURE

The chemical composition of the steel strip used in this study is given in Table I. This type of material is fabricated by a local steelmaker and the obtained samples are in the box annealed condition prior to the temper rolling operation. The thickness of the strip is 0.580 mm and its microstructure consists of equiaxial ferrite grains with an average size of approximately 17 μm. Tensile samples are machined with axes parallel to the strip rolling direction. After that, the deformed samples are air-annealed at 700°C during times ranging from 600 to 7200 seconds.

Table I. Chemical composition of experimental material.

C	S	Si	Mn	Al	Cu	Mo	P	Ni
0.072	0.0012	0.570	0.577	0.211	0.034	0.014	0.042	0.033

The initial, deformation and annealing microstructures are characterized on longitudinal cross-sections using optical microscopy and orientation image microscopy (OIM) based in electron backscattering diffraction, EBSD. In preparation for optical metallography and OIM analysis the samples are polished and subsequently etched in a 2% Nital solution to reveal all the grain boundaries. Average grain size is evaluated using commercial image analysis software based on the linear intercept method. Grain orientation spread (GOS), image quality (IQ) and texture maps are obtained in a Philips XL30-ESEM scanning electron microscope equipped with TSL-OIM V.3 software.

RESULTS

The evolution of grain size during annealing at 700 °C of material deformed 8% in tension is illustrated in Fig. 1. As can be seen, after an incubation period of about 700 seconds, grain growth occurs in two consecutive linear stages: a region of relatively fast linear grain growth leads to a region of very slow grain growth after 3000 seconds of annealing.

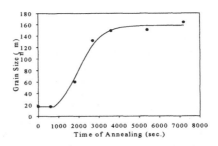

Figure 1. Evolution of grain size during annealing at 700 °C of electrical steel deformed 8% in tension.

The microstructures of deformed and partially recrystallized samples are showed in Fig. 2. The ferrite microstructure of the as-deformed material (Fig. 2a) exhibits no appreciable change in grain size and morphology with respect to the microstructure prior to deformation. However, after annealing during 1800 seconds at 700°C, it is observed that some of the deformed grains grow abnormally with no evidence of primary recrystallization (Fig. 2b).

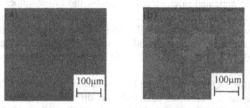

Figure 2. Optical micrographs showing the microstructures of: (a) box annealed electrical steel deformed 8%in tension; (b) partially recrystallized microstructure after annealing during 1800 seconds at 700 °C.

Figure 3 illustrates the $\phi_2=45°$ sections of the orientation distribution functions (ODF) representing the surface textures of samples in the as-received condition (Fig. 3a), after 8% strain in tension (Fig. 3b) and after annealing during 1800 s at 700 °C (Fig. 3c). The texture of the as-received sample can be described by the so-called γ-fiber ({111}<uvw>) and two additional components belonging to the so-called θ-fiber ({100}<uvw>). This type of texture is similar to that usually observed in cold rolled and annealed low carbon steels. The texture after tensile deformation is similar although the θ components appear rotated slightly around the <001>-fiber axis (Fig. 3b). Finally, after annealing, reappearance of the θ components present in the as-received condition and a decrease in the ODF values of the components in the γ-fiber are observed. These results suggest that the abnormal growth of some of the grains (Fig. 2b) in the deformed microstructure occurs at the expense of grains with γ-fiber orientations.

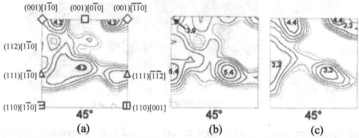

Figure 3. $\phi_2=45°$ sections of the ODF's representing the surface textures of samples in: (a) as-received condition, (b) strained 8% in tension and (c) annealed 1800 s at 700 °C.

Maps of grain orientation spread (GOS), which is a measure of the average misorientation within individual grains in the microstructure, are illustrated in Fig. 4 for the as-received condition (Fig. 4a), after 8% strain in tension (Fig. 4b) and after annealing during 1800 s at 700 °C (Fig. 4c). The grey level scale indicates the values of average misorientation: darker grey level corresponds to larger GOS values. The GOS values for the sample in the as-received condition (Fig. 4a) are very low (global average 0.44). According to Humphreys [8], low values of GOS indicate that the material is fully recrystallized. In contrast, the GOS values for the sample strained to 8% elongation in tension (Fig. 4b) are significantly larger giving a global average of 2.6. This increase in GOS values is associated to the effect on local misorientation caused by the introduction of dislocations during deformation. Finally, as observed in Fig. 4c, the larger grains in the annealed microstructure exhibit the lower values of GOS (similar to those observed in the recrystallized as-received material) and appear to be growing into grains with larger values of grain orientation spread. These observations suggest that, during annealing, primary recrystalization by nucleation and growth of new grains in the deformed microstructure does not take place and the recrystallization in this material occurs by a mechanism of strain – induced boundary migration.

Image quality (IQ) maps indicate the perfection of the crystal lattice in the region of the microstructure where the electron backscattered diffraction patters are obtained. Thus, any distortion in the crystal lattice causes low image quality values [5, 9]. Fig. 5a illustrates the variation of the ratio IQ/GOS for different grain orientations (texture components) located along the so-called α fiber of the ODF for the sample deformed 8% in tension, Fig. 3b. These orientations are located along the $\phi_1=0$ line in the $\phi_2=45°$ section of the ODF. As can be seen, the IQ/GOS ratio decreases rapidly as ϕ increases along this fiber. The EBSD data obtained for this sample is partitioned to illustrate the orientations that have the highest grain average IQ values and, therefore, the lowest lattice distortion in the strained material. The partition is performed for grain orientations with average IQ values larger than 120. Fig. 5b illustrates the $\phi_2=45°$ section of the ODF calculated for this partition. As can be seen, the grain orientations that exhibit the lowest distortion (higher IQ values) after 8% deformation in tension are those belonging to the θ-fiber. A similar procedure is performed to find the grain orientations with the higher GOS values. The results are similar to those shown in Fig. 5b. Therefore, grain orientations with larger IQ values and lower GOS values are along the θ-fiber ($\phi=0$, $\phi_2=45°$) of the ODF of the deformed material (Fig. 3b).

Figure 4. Grain orientation spread (GOS) maps for (a) material in the as-received condition, (b) strained 8% in tension and (c) annealed 1800 s at 700 °C. The GOS values increase as the gray level in the images increases.

DISCUSSION

The results of this investigation show that isothermal annealing of the present electrical steel deformed in 8% tension causes grain growth following a linear kinetics until a stable grain size of about 150 μm is reached. During this process, some selected grains grow at the expense of surrounding grains in the matrix and no evidence of primary recrystallization by nucleation and growth of new grains is observed. These results are similar to those reported by other researchers [10-12]. The driving force for boundary migration is the difference in stored energy between neighboring deformed grains [5-7, 13-15] and the processes is called strain-induced boundary migration (SIBM). This process is favored at small deformations where the stored energy is not large enough to cause nucleation of new grains in the strained microstructure.

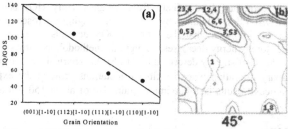

Figure 5. (a) Variation of IQ/GOS for grain orientations located along the α-fiber of the ODF for the sample deformed 8% in tension. (b) $\phi_2=45°$ section of the ODF calculated for a partition of orientations with IQ values greater than 120 in the sample deformed 8% in tension.

During deformation of steel, rearrangement of the dislocation distribution by polygonization, cell and/or subgrain formation in regions of high density of tangled dislocations causes dynamic recovery [15]. The presence of cell or subgrain boundaries causes variations of orientation within the deformed grains. According with Humphreys [8], the local stored energy of deformed grains depends on the degree of misorientation across the cell/subgrain boundaries (θ) and the cell or subgrain diameter (D), see the ecuation 1:

$$E_D \approx \frac{K\theta}{D} \tag{1}$$

where K is a constant. Thus the stored energy decreases as the ratio θ/D decreases. Dillamore et al. [16] and Samet-Meziou et al. [17] measured the size and misorientation between neighboring cells in deformed polycrystalline iron using transmission electron microscopy. Variations in cell size between 0.5 and 1 µm and average misorientation between 2 and 6° are observed depending on the grain orientation. The combination of smaller subgrain size and larger average misorientation is interpreted as an estimation of the local value of stored energy. The results showed that for orientations with [110] parallel to the rolling direction, the local stored energy increases as the crystal plane parallel to the rolling plane changes according to: $E_{\{110\}}>E_{\{111\}}>E_{\{112\}} > E_{\{100\}}$. Figure 3a shows that the (001)[1$\overline{1}$0], (112)[1$\overline{1}$0], (111)[1$\overline{1}$0] and (110)[1$\overline{1}$0] orientations are along the α-fiber ($\phi_1=0$, $\phi=0$-π/2, $\phi_2=45°$) and, according with Fig. 5a, the IQ/GOS ratios for these orientations in the deformed microstructure decrease in the same order as that observed [16, 17] for the local stored energy. Therefore, it can be concluded that orientations with the highest IQ/GOS ratio also exhibit the lowest local stored energy as a result of deformation. As discussed above, the driving force for boundary migration is the difference in stored energy between neighboring grains. In this mechanism, a boundary migrates into the grain with larger local stored energy. Thus, the results of this investigation indicate that, annealing at 700 °C of the present electrical steel strip deformed 8% in tension causes growth of orientations with {001}//RD at the expense of orientations with {111}//RD by a mechanism of strain-induced boundary migration. This process leads to a maximum grain size of about 150 μm after about 3600 seconds of annealing.

CONCLUSION

The results of this investigation show that the local stored energy resulting from deformation may be estimated qualitatively from image quality (IQ) and average grain orientation spread (GOS) values obtained by orientation imaging microscopy in the scanning electron microscope. It is shown that the ratio IQ/GOS depends on grain orientation and increases as the local stored energy decreases. Using this methodology shows that grain growth during annealing at 700 °C of slightly deformed low-C, Si-Al electrical steel strips takes place by strain-induced boundary migration where: θ-fiber orientations grow at the expense of γ-fiber orientations. This process results in a maximum grain size of about 150 μm after about 3600 seconds of annealing.

ACKNOWLEDGMENTS

The authors gratefully acknowledge the technical assistance of Ms. Martha Rivas, Felipe Marquez and the financial support from CONACYT.

REFERENCES

1. S. W. Cheong, E.J. Hilinski, and A.D. Rollett, Metall. Trans 34A, 1311 (2003).
2. J. Barros, J. Schneider, K. Verbeken and Y. Houbaert: J. Magn. Magn. Mater. 320, 2490 (2008).
3. Philip Beckley, *Electrical Steels for Rotating Machines*, first edition (The Institution of Electrical Engineers, London, United Kingdom, 2002) p. 27.
4. A. R. Mader, Metall. Trans. 17A, 1277 (1986).
5. Jongtae Park, Jerzy A. Szpunar and Sangyun Cha, Mater. Sci. Forum Vol. 408-412, 1263 (2002).
6. S. W. Cheong, E.J. Hilinski, and A.D. Rollett, Metall. Trans 34A, 1321 (2003).
7. L. Kestens, J.J. Jonas, P.Van Houtte, and E. Aernoudt, Metall. Trans. 27A, 2347 (1996).
8. F. J. Humphreys, J. Materials Science 36, p.3833, (2001);
9. J.Park, J.A. Szpunar. Acta Materala 51, p. 3037, (2003)
10. R. W. Ashbrook, Jr. and A.R. Mader: Metall. Trans. 16A 897 (1985).
11. C. Antonione, G. Della Gatta, G. Riontino, G. Venturello, J. Mater. Sci. 8, 1 (1973).
12. F. Marino, C. Antonione, G. Riontino, M. C. Tabasso, J. Mater. Sci. 12, 747 (1977).
13. Seung- Hyun Hong and Dong Nyung Lee, Mater. Sci. and Eng. A 375, 75 (2003).
14. F. J. Humphreys, Mater. Sci. Forum Vol. 467-470, 107 (2004).
15. F. J. Humphreys and M. Hartherly: *Recrystallization and Related Annealing Phenomena*, second edition (Elsevier Science, UK 2004) p.248.
16. I.L. Dillamore, C. J. E. Smith and T.W. Watson, Metal Sci. 1, 49, (1967).
17. A. Samet-Meziou , A.L. Etter , T. Baudin, R. Penelle. Mat. Sci. and Eng. 473 342 (2008).

Mater. Res. Soc. Symp. Proc. Vol. 1276 © 2010 Materials Research Society

Evolution of microstructure of 304 stainless steel joined by brazing process

F. García-Vázquez, I. Guzmán-Flores, A. Garza and J. Acevedo
Corporación Mexicana de Investigación en Materiales (COMIMSA), Calle ciencia y
tecnología No. 790, Col. Saltillo 400, C.P. 25290, Coahuila, México
email: felipegarcia@comimsa.com

ABSTRACT

Brazing is a unique method to permanently join a wide range of materials without
oxidation. It has wide commercial application in fabricating components. This paper
discusses results regarding the brazing process of 304 stainless steel. The experimental
brazing is carried out using a nickel-based (Ni-11Cr-3.5Si-2.25B-3.5Fe) filler alloy. In this
process, boron and silicon are incorporated to reduce the melting point, however they form
hard and brittle intermetallic compounds with nickel (eutectic phases) which are
detrimental to the mechanical properties of brazed joints. This investigation deals with the
effects of holding time and brazing temperature on the microstructure of joint and base
metal, intermetallic phases formation within the brazed joint as well as measurement of the
tensile strength . The results show that a maximum tensile strength of 464 MPa is obtained
at 1120°C and 4 h holding time. The shortest holding times will make boron diffuse
insufficiently and generate a great deal of brittle boride components.

INTRODUCTION

High temperature brazing with nickel-based filler metal, produces high performance
joints with excellent load resistance as well as high corrosion resistance [1-3]. The method
has been widely used in high technology industries as a cost effective means and offers a
series of advantages. For instance, brazing can be used to join complicated assemblies
between thick and thin sections, odd shapes, or differing wrought and cast alloys. A single
step treatment is normally necessary to produce integral components with this technique in
a vacuum brazing furnace. For conventional brazing, the joint gaps to be brazed are
generally required to retain about 0-1.5 mm to provide capillary attraction [2]. Furthermore,
one essential property of this process is a strong metallurgical reaction between the brazing
alloy and the base metal that results in joints of high strength and toughness. It process is
employed in the joining and repair of aeroengine hot section components manufactured
from nickel-based superalloys [4]. It is also applied to components that are too badly
damaged for weld repair alone and in situations where welding causes appreciable
mechanical distortion. Using filler alloy containing boron and silicon are known to form
hard and brittle intermetallic phases at the final microstructure of the brazed joint [5].
Formation of eutectic constituents and other second phase particles in a continuously
distributed fashion either along the central region of the joint or at the base metal–braze
interface are often found to be deleterious to the properties of brazed joints [6].

On the other hand AISI 304 stainless steel has excellent corrosion resistance, and
operates within certain temperature ranges. It is used in the manufacture of heat

exchangers, cooling systems and power generators [7]. The main reason for using vacuum brazing process as a joining method of these components is to obtain minimal distortion of the base metal without porosity, besides it can be performed into complex junctions. In brazed joints it is very important to reduce the amount of intermetallic compounds (mainly silicates and borides) in the center of the joined area. Several studies reveal that holding time and brazing temperature can increase the dissolution of intermetallic compounds in the solid solution by isothermal solidification [9].

EXPERIMENTAL PROCEDURE

In the present research, 304 stainless steel coupons are used to evaluate microstructure features and tensile strength of brazed joints. Brazing is carried out in a vacuum furnace using a nickel-based filler alloy at temperatures of 1120°C and 1160°C for 2 and 4 hours holding time. Specimens for metallographic examination are cut normal to the brazed joint from each sample. To reveal the joint microstructure, cross-sections of the joint are ground, polished, and etched with a solution of 10 ml H_3PO_4 + 50ml H_2SO_4 + 40ml HNO_3 + 100mL H_2O. Microstructural characterization of the brazed joints is performed by using a JEOL scanning electron microscope (SEM) equipped with EDXS semi quantitative analysis using an accelerating voltage of 15 kV, spot size of 40 and a working distance of 11 mm. In order to evaluate the tensile strength of the brazed joints, 304 stainless steel specimens are prepared according to AWS C3.2M and ASTM E8 standards. Tension test is realized in a 100 ton Olsen Tensile machine.

RESULTS AND DISCUSSION

Figure 1 shows the cross section of the brazed joints microstructure obtained with 1120°C and 1160°C for 2 and 4 hours. As it can be appreciated diffusion zone or Ni solid solution area increases by a raise of brazing holding time. This behavior can be also observed with an increase of brazing temperature. Brazing temperature and holding time have a remarkable effect on intermetallic compound formation at the joint center.

Figure 2 illustrates the brazed joint microstructure and consists of the diffusion zone (DZ) or isothermal solidification zone, the athermally solidification zone (ASZ) and the interface reaction (IR). As it can be seen in Figure 2b and 2d, the EDXS spectrum corresponding to (DZ) presents a high content of Ni, Fe, Cr and Si due to Fe interdiffusion towards filler metal. In ASZ zone, different intermetallic compounds Fe-Si, Fe-C-Si, Ni-Si and Ni-Fe are formed. Boron is used to depress the melting point of nickel-based filler alloys, and the insufficient diffusion of this element from the brazed zone into the base material results in liquid phase along the joint zone [9,10].

The microstructures in the athermally solidified zone are silicides and chromium borides formed by diffusion of boron and chromium. This area is formed because there is not enough time for isothermal solidification. Figure 3a shows the intermetallic compounds formation and the EDXS spectrum illustrates the formation of different intermetallic compounds such as Ni-Fe-Si and Ni-Fe-Cr, Figure 3b. These intermetallic compounds are found to be eutectic type and they are brittle and very hard.

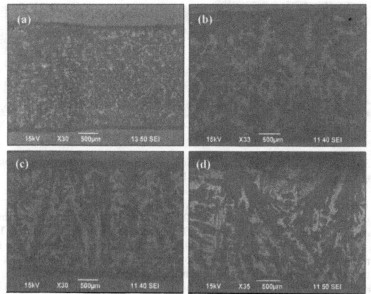

Figure 1. Cross section microstructure of brazed joints of 304 stainless steel with nickel-based filler alloy: (a) 1120°C 2h, (b) 1120°C 4h. (c) 1160°C 2h and (d) 1160°C 4 hours.

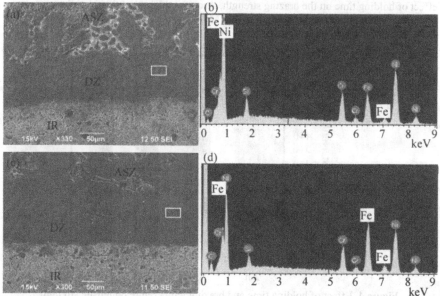

Figure 2. Secondary electron image of the microstructure of diffusion zone at the interface base metal/filler material, and EDXS spectrum corresponding to DZ microarea. (a) and (b) 1120°C for 2 h. (c) and (d) 1120°C for 4 h.

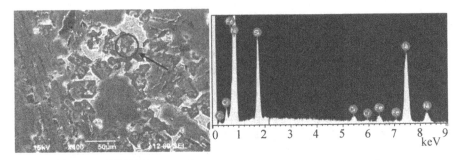

Figure 3. (a) Ni-Fe-Si intermetallic compound formation in the center of the joint, (b) EDXS spectrum of the intermetallic indicated by the arrow in (a).

The results of tensile strength are shown in Figure 4, as it can be seen that a maximum tensile strength is obtained with 1120°C and 4h. The holding time at brazing temperature has a great effect on the boron diffusion and the base metal dissolution. To achieve a proper brazing microstructure and higher tensile strength, the holding time should be optimized. Increasing the holding time below a critical holding time can make boron diffuse to base metal adequately by a homogeneous microstructure formation which increases the strength greatly. Increasing the holding time beyond a critical time will lead to the over-dissolution of base metal, which results in the generation of corrosion voids and decreases the strength. In this paper, an experimental method is performed to discuss the effect of holding time on the brazing strength, through which an optimal holding time is obtained. A maximum tensile strength of 464 MPa is obtained with brazing temperature of 1120°C and 4 hours holding time.

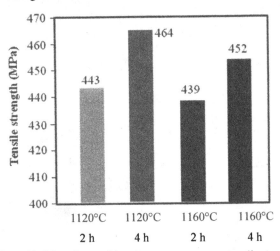

Figure 4. Effect of holding time and brazing temperature on tensile strength.

CONCLUSIONS

The microstructure at the joint area shows 3 zones consisting of isothermal solidification zone or diffusion zone (DZ) formed by base Ni solid solution, an athermal solidification zone (ASZ) which intermetallic compounds formations occurs and interface reaction (IR). The diffusion of boron and silicon increases by increasing brazing holding time, therefore under these conditions a suitable intermetallic compounds formation is maintained and the diffusion zone is thicker. Too short holding time will make Boron diffuse insufficient and generate a great deal of brittle Boride components, and too long holding time will make the base metal dissolve into the filler metal excessively and create more corrosion voids. For a time of 4 hours and 1120°C a maximum tensile strength of 464 MPa is achieved.

REFERENCES

1. Y.H. Yu, M.O. Lai. Effect of gap filler and brazing temperature on fracture and fatigue of wide-gap brazed joints. J. Mater. Sci. 30 (1995) 2101-2107.
2. E. Lugscheider, Th. Schittny, E. Halmoy. Metallurgical aspects of additive-aided wide-clearance brazing with nickel-based filler metals. Welding journal (1989) 9s-13s.
3. W.D. Zhung, T.W. Eagar. Transient liquid phase bonding using coated metal powders. Welding journal (1997) 157s-162s.
4. S.K. Tung, L.C. Lim. Wide gap brazing with precraks of nickel base braze mixes. Mater. Sci. Technol. 11 (1995) 949-954.
5. R. Johnson, M. Baron and N.J. Livesey, Third International Brazing and Soldering Conference (BABS), Paper 21, 1979.
6. E. Lugsscheider, H. Schmoor, U. Eritt, *Brazing, High Temperature Brazing and Diffusion Welding*, Deutscher Verlag fur Schweisstechnik GmbH, Germany, 259–261 (1995).
7. F. Pra, P. Tochon, Ch. Mauget, J. Fokkens and S. Willemsen. Promising designs of compact heat exchangers for modular HTRs using the Brayton cycle. Nuc. Eng. Des. 2008; 238(11): 3160-3173.
8. Wenchun Jiang, Jiangming Gong, Shan-Tung Tu, Hu Chen. Effect of geometric conditions on residual stress of brazed stainless steel plate-fin structure. Nuc. Eng. Des. 2008; 238 (7): 1497-1502.
9. W.F. Gale, E.R. Wallach. Microstructural development in transient liquid-phase bonding. Metall. Trans. A 22 (1991) 2451–2457.
10. R.D. Eng, E.J. Ryan and J.R. Doyle. Solidification phenomena in nickel base brazes containing boron and silicon. Welding Journal, 56, 15 (1977).

Mater. Res. Soc. Symp. Proc. Vol. 1276 © 2010 Materials Research Society

Fourier Thermal Analysis of Eutectic Al-Si Alloy with Different Sr Content

R. Aparicio [1], G. Barrera [2], G. Trapaga [1] and C. Gonzalez [3]
[1]CINVESTAV Queretaro, Libramiento Norponiente 2000, Real de Juriquilla, 76230 Queretaro Qro. Mexico, raparicio@qro.cinvestav.mx
[2]Instituto de Investigaciones Metalúrgicas, UMSNH. Apdo. Postal 888 Centro, 58000. Morelia Mich. México. gbarrera@umich.mx
[3]Department of Metallurgical Engineering, Facultad de Química, UNAM, Edificio "D" Circuito de los Institutos s/n, Cd. Universitaria, 04510 México D. F., México. carlosgr@servidor.unam.mx

ABSTRACT

The purpose of this work is to explore the capability of Fourier Thermal Analysis (FTA) to detect differences in solidification kinetics between unmodified and Sr modified eutectic Al-Si alloy obtained from the same base alloy. Experimental melts are produced in silicon carbide crucibles using an electrical resistance furnace and burdens of A356 alloy and commercial purity Si. The addition of strontium to the melts is accomplished using Al-10 pct Sr master alloy rod. Chemical composition is controlled using spark emission spectrometry. The changes in microstructure are characterized using optical microscopy. Thermal analysis are performed in cylindrical stainless steel cups coated with a thin layer of boron nitride, using two type-K thermocouples connected to a data acquisition system. Experimental cooling curves are numerically processed using FTA. Results show changes in solidification kinetics of eutectic Al-Si alloy with different Sr content. These changes, measured at the beginning and during solidification of the probes, can be related to the changes in nucleation and growth causing the differences detected during microstructural characterization of the probes.

INTRODUCTION

For commercial Al-Si cast alloys it is a common practice during melt treatment to include the addition of little quantities of strontium in order to refine the coarse flake-like structure of the silicon phase of the eutectic constituent to a fine fibrous structure [1], improving properties of cast components.

There are two main theories trying to explain eutectic modification. The first theory is focused on changes of the growth of eutectic silicon by a mechanism named impurity induced twinning [2]. The second theory is based in an inhibition of nucleation of eutectic silicon that results in more undercooling and the refinement of microstructure [3, 4].The real mechanism of modification is probably related both to the reduction in nucleation and the changing growth of eutectic silicon by impurity induced twinning. However it is still not yet understood [5].

Taking into account that modification mechanisms have been explained in terms of changes in nucleation and growth of eutectic silicon it is interesting to explore the changes in solidification kinetics of Al-Si alloys as a result of addition of eutectic modifiers. Computer aided cooling curve analysis (CA-CCA) methods have been used to study solidification kinetics of various alloy systems of metallurgical interest [6-8]. Newton thermal analysis (NTA) and Fourier thermal analysis (FTA) are the most representative techniques. It has been found that FTA is the most reliable method because it takes into account the presence of thermal gradients in the probe using data acquired from two thermocouples located at two different radial positions within the sample to obtain the zero baseline curve. Fundamentals, limitations, and implementation of this method have been discussed elsewhere [9-11].

The purpose of this work is to explore the capability of Fourier Thermal Analysis (FTA) to detect differences in solidification kinetics between unmodified and Sr modified eutectic Al-Si alloy.

EXPERIMENT

In order to obtain a near eutectic Al-Si alloy, melts are produced in an electric furnace with an argon atmosphere using burdens of commercial purity silicon and A356 alloy. Chemical composition is adjusted using spark emission spectroscopy. Table I shows chemical composition of the experimental base melt. The addition of strontium is accomplished using Al-10%Sr master alloy. The experimental alloys showed strontium levels from 2 to 210 ppm.

Table I. Chemical composition of experimental base melt prior to modification.

	Si	Fe	Cu	Mn	Mg	Zn	Ni	Sr	Al
Al-Si	12.57	0.4746	0.016	0.1063	0.23	0.0166	0.005	0.0002	Bal

Thermal analysis test samples are taken by submerging a cylindrical stainless steel test cup (0.03 m inner diameter, 0.05m in height and 1.5 mm in thickness, covered with boron nitride) into the melt. The cups are kept submerged for approximately 30 seconds to allow them to reach the bath temperature. Then, they are removed from the melt and placed on a thermal analysis test stand where they are isolated thermally at the top and the bottom. A compressed air ring is used in order to study the effects of a high cooling rate on cooling curves, microstructure and solidification kinetics characterization of experimental probes. The air ring is surrounding the mold containing the sample. In experiments of high level of cooling rate, a flow of 77 liters per minute of air is used. For the low cooling rate condition, natural room air cooling is allowed. In order to record the thermal history of the alloy in the thermal analysis test stand during cooling, two thermocouples type K, with alumina sheath, 0.0015 m OD, are introduced in the liquid sample at the same depth at two different radial positions. Thermocouples output is recorded in a personal computer connected to a NI Field Point cFP 1804 data acquisition system. A calibration procedure is performed with 99.9 % aluminum. The experimental cooling curves are numerically processed using FTA method in order to obtain information about the evolution of solid fraction during solidification of the sample.

For microstructural analysis, the samples are sectioned transversally and prepared by standard polishing procedures. The microstructure of the specimens is observed using optical microscopy.

RESULTS AND DISCUSSION

Figure 1 shows typical cooling curves of experimental Al-Si alloys with different Sr contents. It can be seen in Fig. 1(a) that the cooling and solidification process of the alloys under study can be divided in three main steps. The first step is the cooling of the liquid where temperature drops continuously from the initial temperature to the start of the eutectic plateau. At this time the second stage, corresponding to the eutectic solidification begins. The third step corresponding to the cooling of the solid begins at the end of eutectic plateau where temperature drops again as a result of the cooling of the fully solid sample. It can be seen in this figure that there are displacements down and to the right of the cooling curves as Sr content is increased

In Fig. 1(b) it is shown in more detail the effect of the presence of Sr on the position of the eutectic plateau and cooling curves. It can be observed that, during eutectic solidification, an increase in Sr content promotes the increase of the acting undercooling, defined as the difference between equilibrium eutectic temperature (577 °C) and measured operating temperature during solid formation.

This effect has been associated to an increase in the nucleation barriers [4] and a decrease in the growth ability of eutectic Si, increasing twin density and thus the kinetic undercooling [2].

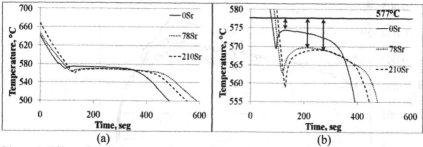

(a)   (b)

Figure 1. Effect of strontium content on cooling curves.

Fig. 2 shows the microstructure observed in experimental probes associated to the cooling curves shown in Fig. 1. In Sr-free alloys, Fig. 2(a), it can be noticed the presence of polyhedral crystals and primary sharp silicon into an aluminum matrix. Under the presence of a Sr content of 77 ppm, figure 3(b), these primary silicon crystals disappear leaving a less acicular form of eutectic Si and the presence of some primary phase aluminum dendrites. Finally in micrograph of figure 3c it can be observed eutectic silicon totally modified with the presence of Al-rich dendrites as the white phase. Fig. 2 suggest that an increase in Sr content eliminates silicon primary crystals originally present in the Sr-free eutectic alloy and promotes the gradual modification of the eutectic and the presence of dendrites of Al-rich primary phase.

The experimental cooling curves are numerically processed according to FTA procedure in order to obtain the solidification kinetics of the eutectic alloy with different Sr contents and the results are shown in Fig. 3.

A plot of solid rate formation against solid fraction is commonly used to describe solidification kinetics because this representation allows the analysis of solidification rate evolution as the micro constituents nucleates and grows. The solidification rate evolutions of the eutectic alloy with different Sr content depicted in Fig.3 show important differences during the beginning of solidification, i.e. nucleation and also in the presence of higher Fs values , i.e., during growth.

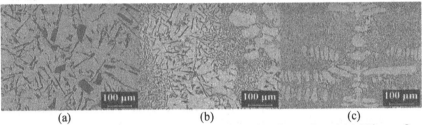

(a)   (b)   (c)

Figure 2. Micrographs of (a) unmodified Al-Si alloy, (b) 78 ppm Sr and (c) 210 ppm Sr.

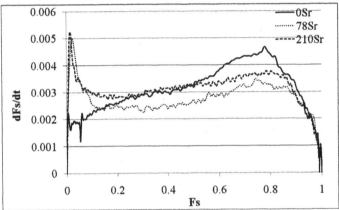

Figure 3. Effect of strontium on FTA solidification kinetics of experimental alloy at low cooling rate condition.

It can be seen that, as a result of a Sr increase, at low values of Fs, during eutectic nucleation stage, there is an increase in the initial solidification rate that may be associated to the poisoning of the nucleation sites for eutectic silicon and to the presence of dendrites of primary Al- rich phase. Apparently the presence of Sr difficult   eutectic nucleation as suggested by the elimination of the primary silicon crystals originally present in the Sr free alloy of Fig. 2(a) and promotes the growth of dendrites of primary phase, as suggested by microstructural results. The increase the amount of primary phase formed during solidification can be the result of an increase of kinetic barriers for the eutectic Al-Si nucleation.

It is observed in Fig. 3 that after first stages of solidification, i.e., during eutectic growth, FTA results shows that fully and partially modified eutectic grows at lower velocities than the velocities present in the unmodified, strontium free sample

Fig. 4 shows the effect of a higher cooling rate on the cooling curves and solidification kinetics of the Sr free eutectic alloy. It can be seen in this figure that an increase in cooling rate under experimental conditions used in this work produce an increase in the operating undercooling of approximately 5°C during eutectic solidification, and an unambiguous increase of the solidification rate. Fig. 5 shows the microstructures of the near eutectic alloy without Sr solidified under the two experimental cooling conditions. It can be seen that an increase in cooling rate under experimental conditions causes the refinement of microstructure which is apparently associated to the increase in solidification rate caused by the increase in operating undercooling.

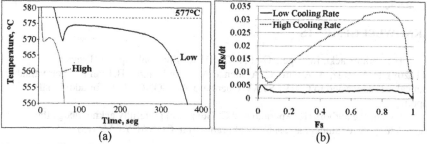

(a)   (b)

Figure 4. Effect of cooling rate on (a) cooling curves and (b) FTA solidification kinetics of Sr -free eutectic Al-Si alloy.

(a)   (b)

Figure 5. Effect of cooling rate on microstructure of near eutectic Al-Si alloy without Sr: (a) Low cooling rate and (b) High cooling rate.

It is interesting to notice that Fig.4 associated to an increase in cooling rate shows that an increase in operating undercooling of about 5 °C produce an unambiguous increase in eutectic solidification rate while the undercooling increase caused by the Sr content increase shown in Fig. 1(b), of similar magnitude, shows relatively little effect on changes in solid rate formation, see Fig. 3. This behavior suggest that the solidification mechanism acting during Sr modified eutectic solidification, under the presence of an increase of the kinetic component of eutectic undercooling is different that the mechanism operating in the Sr free eutectic alloy.

CONCLUSION

FTA method has been able to detect changes in the eutectic solidification kinetics Al-Si alloys as a result of the presence of different amounts of Sr causing different levels of eutectic modification.

FTA results of samples with different Sr contents show important differences in solidification rate evolution at the beginning of solidification suggesting that Sr restrain the eutectic Si nucleation ability originally present in the unmodified alloy, which rends difficult the start of eutectic growth and enhances the formation of dendrites of primary phase.

During eutectic growth, FTA results shows that fully and partially modified eutectic grows at slightly lower velocities than the velocities present in the unmodified, strontium free sample.

ACKNOWLEDGEMENTS

The authors acknowledge DGAPA UNAM for financial support (Project IN112209), to A. Ruiz, A. Amaro, C. Atlatenco A. Galindo and I. Beltran for their valuable technical assistance and to CINVESTAV Queretaro and CONCYTEQ for additional support.

Two of the authors (R. Aparicio and C. Gonzalez) gratefully acknowledge the financial support for student and sabbatical grants from CONACYT Mexico.

REFERENCES

1. J. E. Gruzleski: *American Foundrymens Society* Inc. (1990).
2. S.Z. Lu, A. Hellawell: *Metall. Trans. A*, Vol. 18A (1987), p. 1721.
3. A.K. Dahle, K. Nogita, J.W. Zindel, S.D. McDonald, and L.M. Hogan: *Met. Mater. Trans. A*, Vol. 32A (2001), p. 949.
4. Y.H.Cho, H.C.Lee, K.H. OH and A. K. Dahle: *Met. Mater. Trans. A*, Vol. 39A (2008), p. 2435.
5. Y.H. Cho and A.K. Dahle: *Met. Mat. Trans. A*, Vol. 40A (2009), p. 1011.
6. C. Gonzalez-Rivera, B. Campillo, M Castro, M. Herrera, J. Juarez-Islas: *Materials Science and Engineering A*, Vol. 279 (2000), p. 149.
7. González R. Carlos, Cruz M. Héctor, García H. José, Juarez I: *J. Mater. Eng. Performance*, Vol. 8.1 (1999), p. 103.
8. A.Cetin and A. Kalkanli: *J. Mater. Proc. Tech.*, Vol. 209 (2009), p.4795.
9. W. Kapturkiewicz, A. Burbielko, and H.F. Lopez: *AFS Trans.*, Vol. 101 (1993), p. 505.
10. J.O. Barlow and D.M. Stefanescu: *AFS Trans.*, Vol. 105 (1997), p. 339.
11. D. Emadi and L. Whiting: *AFS Trans.*, Vol. 110 (2002), p. 285.

Mater. Res. Soc. Symp. Proc. Vol. 1276 © 2010 Materials Research Society

Synthesis of nanostructured metal (Fe, Al)-C_{60} composites

I. I. Santana García[1], V. Garibay Febles[2], H.A. Calderon[1]

[1]ESFM, Instituto Politécnico Nacional, Edif. 9 UPALM D.F. 07738, México

[2]LMEUAR, Instituto Mexicano del Petróleo, México D.F. 07730, México

ABSTRACT

Composites of M-2.5 mol. % Fullerene C_{60} composites (where M= Fe or Al) are prepared by mechanical milling and Spark Plasma Sintering (SPS). The SPS technique has been used to consolidate the resulting powders and preserve the massive nanostructure. Results of X-Ray Diffraction and Raman Spectroscopy show that larger milling balls (9.6 mm in diameter) produce transformation of the fullerene phase during mechanical milling. Alternatively smaller milling balls (4.9 mm in diameter) allow retention of the fullerene phase. SEM shows homogeneous powders with different particle sizes depending on milling times. Sintering produces nanostructured composite materials with different reinforcing phases including C_{60} fullerenes, diamonds and metal carbides. The presence of each phase depends characteristically on the energy input during milling. Transmission Electron Microscopy (TEM) and Raman Spectroscopy show evidence of the spatial distribution and nature of phases. Diamonds and carbides can be identified for the sintered Fe containing composites with a relatively high volume fraction.

INTRODUCTION

Fullerene phases are of great interest due to their diversity and attractive mechanical and physical properties. Since their discovery, a great deal of attention has been given to their characterization and potential application. A fullerene composite comprises a dispersion of ultrafine particles of fullerene incorporated in a matrix, representing a new class of technologically relevant composites. Recently, the development of fullerene-reinforced composite materials inside a metallic matrix is directed towards the improvement of mechanical properties [1-4].

Mechanical alloying is a very popular method to fabricate materials with novel structures and/or properties. It can be used to produce non equilibrium structures, supersaturated solid solutions, metastable crystalline phases and nanocrystalline materials [5]. Massive nanostructured materials are possible because the process involves repeated deformation, fracture and welding of powder particles. Milling of powders involves a considerable number of impacts, resulting in solid state alloying with a controlled nanostructure [6-7]. In the present case, the selected processing can also promote phase transformations of the involved components so that harder disperse phases can be developed with a corresponding improvement in the mechanical properties of the resulting product. Additionally and to preserve the massive nanostructure developed during milling, a special sintering methods is required. The Spark Plasma Sintering (SPS) technique offers a relatively faster processing that produces a uniformly dense material with very limited grain growth. It makes use of temperature, applied pressures and an electric pulse current passing through the sintering mix. The process normally preserves the nanostructure developed during milling. In this study, the fabrication of Metal-Fullerene C_{60}

composites by milling and SPS has been investigated. The purpose of this work is to produce dense and fine grain size specimens.

EXPERIMENT

Mechanical milling in an inert atmosphere (Ar) has been used in order to produce alloyed powders of Fullerene C_{60} (99.99% purity) and an Fe matrix (99.9% purity and an average particle size of 50 μm) and C_{60} into Al (99.9% and 30μm). These materials are used to prepare Fe and Al base composites with 2.5% mol C_{60}. Milling has been varied from 0.5 to 2 hours in order to follow the alloying process. Pure powders have been mixed and milled in a high energy SPEX mill with a ball to powder weight ratio of 8:1 and two different ball sizes, 4.9 and 9.6 mm in diameter. Consolidation of powders has been performed by SPS. Cylindrical solid samples of 20 mm in diameter and 4 mm of thickness are produced by sintering at 775 K for the system Fe-fullerene and 650 K for the Al-fullerene composites with an external pressure equal to 15 kN.

Microstructural characterization is done by means of X-Ray Diffraction (XRD) with Co K_{α} radiation, Scanning Electron Microscopy (SEM), Transmission Electron Microscopy (TEM) and Raman Spectroscopy.

RESULTS AND DISCUSSION

Mechanically milled powders

Mechanical milling of Fe or Al with fullerene C_{60} powders has been applied from 15 min to 2 h in order to monitor the metal-fullerene milling process. Fig. 1 shows XRD patterns taken from as-milled powders of M-C_{60} but using two different ball diameters (4.6 and 9.6 mm). For both metallic matrices, the fullerene structure is clearly retained during mechanical milling regardless of the milling time, when using smaller balls (4.6 mm) as given in Fig. 1b. On the other hand, by using large balls (9.6 mm), the energy input is apparently large enough to promote either a higher comminuting of the fullerenes (making the diffraction peaks broader) or a phase transformation or both. As a result, the diffraction pattern (see Fig. 1a) shows no apparent peaks of C_{60}. However, new peaks are found with a low intensity but clear enough to be seen. They correspond most likely to FeC_3 and diamonds and pointed out by arrows.

Fig. 2 shows SEM images of the Fe-Fullerene C_{60} powders after milling for 30 min and 2 h, respectively, with small milling means (4.6 mm). As expected powder particles are comminuted as a function of milling time. The powder particle size varies from 8 to 20 μm with an average grain size of ~14 μm after 30 minutes of milling. The morphologies of milled Fe-C_{60} powders are rather irregular. This indicates the preponderance of fracture of powder particles at this early milling time and that the fullerene is most likely already mixed in the ductile Fe powder particles. The appearance of relatively large particles most likely means that mixing of Fe and C_{60} has progressed fast with the corresponding agglomeration but still with some possibility to comminute further. Milling for a longer time (2 h) shows smaller particle sizes, as expected for a further fragmentation of powders and the corresponding energy input. The average particle size decreases to 7 μm but this is probably an overestimate since the imaging with an SEM necessarily reduces the total number of small particles in the field of view due to overlapping, projection effects, etc.. Additionally, there is a particular development of the morphology i.e., lamellar particles become predominant over the irregularly shaped large particles found at earlier milling times. However the milling process is most likely following an expected route and accordingly the processing parameters have been correctly selected including the use and amount of control agent (methanol) with very limited agglomeration.

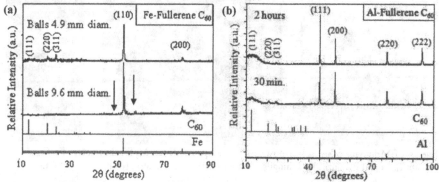

Fig. 1- XRD patterns (a) Fe-C_{60} and (b) Al- C_{60} powder milled at different times with milling balls of 4.6 mm in diameter.

Fig 2- SEM images Fe-C_{60} powder milled at different times using milling balls of 4.6 mm in diameter during (a) 30 minutes and (b) 2 h.

The continued fracture and cold-welding promoted by mechanical milling ultimately result in a homogeneous distribution of the fullerene particles in the Fe or Al metallic matrix. Fig. 3 shows TEM bright and dark field images taken from of Fe-C_{60} powder particles after milling and the corresponding diffraction pattern. The nanostructured grain size distribution can be easily recognized in the powder particles under investigation (Fig. 3a). The crystallite average size is estimated to be approximately 20 nm in the Fe-base materials after 2 h of milling. Additionally, the dark field image (DF-TEM) in Fig. 3b shows that the fullerene is finely and completely dispersed inside the metallic matrix, there is a homogeneous distribution of bright areas. The DF image has been taken with a fullerene reflection from the pattern in Fig. 3c. In all cases the images correspond to a material milled for 2 h with balls of 4.9 mm in diameter. The diffraction patterns in Figs. 3c-d have been taken in convergent beam mode over specific areas, the characteristic fullerene C_{60} hexagonal pattern (corresponding to an fcc arrangement) is clearly seen and its corresponding indexing is given in Fig. 3d.

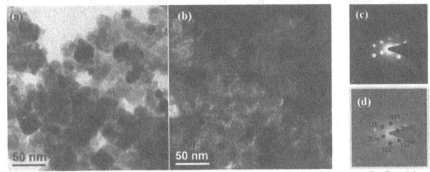

Fig. 3- TEM images of as milled powders in the Spex mill for 2 h for the system Fe-C_{60}. (a) bright field and (b) dark field images (c) microdiffraction pattern, (d) indexing of pattern.

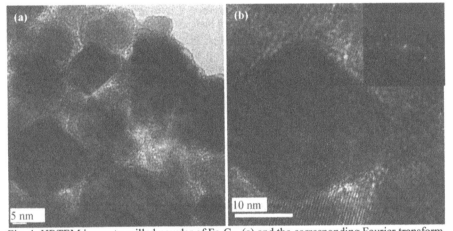

Fig. 4- HRTEM image to milled powder of Fe-C_{60} (a) and the corresponding Fourier transform.

As discussed above (Fig. 1), larger milling balls (9.6 mm) can promote phase transformations and/or comminuting. Figure 4 confirms the existence of nanocrystalline rhombohedral diamond in the as milled Fe-C_{60} powder particles. Figures 4a and b show two views and the lattice image of an specific crystallite in the powder particles, its rhomboidal shape is very apparent and the corresponding lattice parameter (0.251 nm) aggress well with the expected value for a diamond phase[10]. On the other hand, the Fourier transform (insert in Fig. 4b) agrees completely well with such a phase. Other experimental techniques reveal the presence of diamonds in the as milled powders. Raman spectroscopy reveals traces of a diamond phase in a size within the

100

nanoscale. The expected Raman shift Raman for nanocrystalline diamond appears around 1332 cm^{-1} with a wide peak and relatively lower intensity as it is found for these materials [2, 11, 12].

Consolidated materials

Figure 5 shows a typical XRD pattern for the sintered Fe-C$_{60}$ composite. In this case, the identified phases are rhombohedral diamond, iron carbides, iron and C$_{60}$ fullerenes. This material has been milled for 2 h with a low energy input (balls with 4.6 mm in diameter). Apparently C from the fullerenes reacts with the Fe matrix to form the stable cementite phase. Since C$_{60}$ is also present in the solid composites, it is likely that the reaction between C and Fe is preceded by another reaction, most likely leading to amorphous C. As for the diamond phase, it is interesting to discuss the relatively high intensity of the corresponding peaks suggesting a high volume fraction. By comparing this result to that in Fig. 1a, the sintering process has apparently promoted the formation of diamonds.

Fig. 6 shows TEM lattice images taken from a sintered sample. Fig 6a shows a larger section of the sample and there is a Fourier transform for the enlarged image in Fig. 6b. A mosaic of different nanocrystals can be seen with a variety of phases and orientations. The Fourier transform in Fig. 6a shows incomplete rings of reflections at a given lattice parameter distance indicating such a mosaic. The phases are cubic with two extreme cases, relatively large lattice spacing inside another phase with smaller lattice distances. The global Fourier transform shows all this but it gives little indication of the corresponding location in the image. However, this can be achieved to show the phase distribution by amplifying the observation area as done in Fig. 6b. In this image the mosaic is formed by mostly nanosized diamonds but also some areas occupied by fullerenes (see arrows) can be distinguished. The arrows in the Fourier transform indicate the reflections produced by the fullerene structure.

Fig. 5- Diffraction patterns of sintered Fe-C$_{60}$ composite after 2 h of mechanical milling.

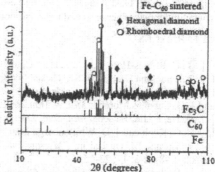

Measurements of Vickers micro-hardness show that the Fe-C$_{60}$ composites reach 128 μHV for composites milled for 2 h [2]. This represents an increase of 28% as compared to the value for pure Fe (100 μHV). On the other hand, porosity is reduced in samples with longer milling times. The smaller powder particle size is most likely responsible for the better packing behavior and lower porosity.

CONCLUSIONS

The present results show that fullerene can transform in the presence of metals, particularly Fe, during mechanical milling. Phase transformations to carbide and diamond phases are induced as a function of input energy by milling and also by the SPS technique. The fullerene phase is stable during milling when the energy input is low (milling balls with diameter of 4.9 mm). High energy input promotes phase transformations to carbides and other phases (diamonds) in relatively high volume fractions and very rapidly. In the system $Fe-C_{60}$ sintered samples show a dispersion of diamonds, carbides and fullerenes in the metallic matrix. Transformation to diamonds is well supported by experimental evidence as a function of milling energy in both as milled and as sintered products. The SPS technique promotes formation of diamonds.

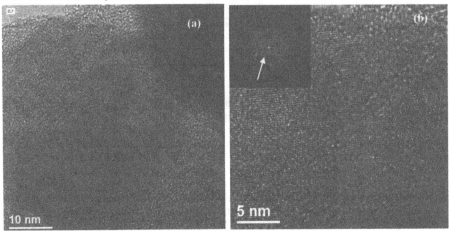

Figure 6. (a) HRTEM image of sintered $Fe-C_{60}$ composite. (b) Mosaic massive nanostructure containing mostly diamonds and the corresponding Fourier transform.

REFERENCES

[1] F. C. Robles-Hernández and H.A. Calderon, *JOM*, **62 -02,** 63-68 (2010).
[2] I. I. Santana García, M. Sc. Thesis, Instituto Politécnico Nacional , 2010.
[3] V.Garibay-Febles, H.A Calderon., F. C. Robles Hernández, M. Umemoto, K. Masuyama and J. G. Cabañas Moreno; *Mat. & Man. Proc.*, **15-4,** 547-567 (2000).
[4] Q. He, C. Jia, J. Meng, *Mat. Scie. Eng.* **A 428,** 314–318(2006).
[5] C. C. Koch and J. D. Whittenberg., *Inter. Comp.*, **4,** 339-355 (1995).
[6] L. Lu and M. On Lai in, *"Mechanical Alloying"*, edited by Kluwer Academic Publishers, USA, (1998), 1-3; 17; 23-65.
[7] C. Surinarayama in, *Mechanical Alloying and Milling,* edited by Marcel Dekker Pub., USA, (2004), pp. 11-17.
[8] M. Tokita, *J. Soc. Pow. Tech. Japan* , **30-11,** 790-804 (1993).
[9] W. Chen, U. Anselmi-Tamburini, J.E. Garay, et al., *Mater. Sci. Eng.* **A-394** , 132 (2005).
[10] ICDD (International Center of Diffraction Data) Identification 79-1479; 26-1075.
[11] F. Tuinstra, and J. L. Koenig, *J. Chem. Phys.*, **53,** 1126 (1970).
[12] F.A. Khalid, O. Beffort, U.E. Klotz, B.A. Keller, P. Gasser, *Diam Rel. Mat.*, **13,** 393-400 (2004).

Mater. Res. Soc. Symp. Proc. Vol. 1276 © 2010 Materials Research Society

Metal-Graphite Couples Synthesized by Means of Mechanical Milling.

I. Estrada-Guel, C. Carreño-Gallardo, R. Pérez-Bustamante, J.M. Herrera-Ramírez and R. Martínez-Sánchez.

Centro de Investigación en Materiales Avanzados (CIMAV), Laboratorio Nacional de Nanotecnología, Miguel de Cervantes No. 120, C.P. 31109, Chihuahua, Chih., México.

ABSTRACT

The aim of this work is the characterization of some graphite-metal couples prepared by mechanical milling (MM). The morphological and microstructural changes during MM of graphite processed with metallic powders of Cu, Ni and Ag (10 and 15 at. %) are studied. Milling is performed in a high-energy ball mill under an inert atmosphere during 1, 4 and 8 hours. The process is also repeated with a pure graphite sample in order to compare the role of metal type and concentration on the morphological characteristics of milled samples. The results show that increasing the concentration of metal particles accelerates the milling process as a result of faster work hardening and particle fracture. The results of X-ray diffraction analysis show that some crystallographic characteristics of the milled couples change as a function of milling time and metal addition. Also, SEM-EDS studies show an important effect of milling time on metal particle distribution in the prepared graphite couples.

INTRODUCTION

Industrial applications utilizing lightweight materials that reduce CO_2 gas emissions and save on energy consumption are an environmentally friendly approach to design [1]. Currently lightweight alloys are used in small quantities due to their relatively low strength which limits their potential applications. To overcome the limitations, extensive research has been done on producing metal matrix composites (MMCs) [2]; these materials have the ability to blend the properties of ceramics (high strength and modulus) with the ductility and toughness of metals [3]. Besides, MMCs offer superior operating performance and resistance to wear [4]. Due its superior properties have been used in the aircraft, space, defense and automotive industries [5].

Al-alloys are some of the most widely used materials as the matrix in MMCs, in industrial applications due to their low density and cost compared with Mg or Ti [6]. On the other hand, graphite fibers have been recognized as high strength, low density materials because of their high strength to mass ratio [7]. Its excellent structural stability and mechanical performance at high temperatures are other reasons to consider graphite as a promissory material [8]. Generally, the processes used to synthesize the Al/C composites can be classified into three categories: (i) liquid phase; (ii) solid phase; and (iii) two phase (solid–liquid) routes [9]. Although the preparation of such composites by liquid routes is by far the most economical [4] there are some inherent problems related both to the apparent non-wettability of graphite by liquid Al-alloys [10] and the large density differences between C and Al. These differences result in porosity formation at the interface [9] that negatively influences the properties of the composites [11]. Therefore, manufacture of uniform dispersions of metal particles in a graphite particulate matrix is an exceedingly difficult task [12]. The distribution of the reinforcement particles depends on the processing route involved. A decrease of the reinforcement particle size can bring about an increase in both mechanical strength and ductility of the composite [13]. Thus efforts have been focused on finding effective dispersion techniques that can disperse the reinforcement particles homogeneously within the matrix [16].

In MMCs fabrication by conventional solid state routes, the segregation of particles are the primary problem due to the different flow characteristic of powders, the agglomeration (to minimize surface energy) and density differences [9]. Furthermore, clusters formed due to non-uniform particle distributions usually result in porosity that reduces the resistance of the material to crack nucleation [14]. Mechanical milling is a method of introducing the reinforcement particles which produces improved particle distributions in the consolidated material [13,15]. As a solid-state technique is performed at low temperatures and facilitates the incorporation of ceramic particles into a metallic matrix [16]. The central event is that the powder particles are trapped between the colliding balls during milling and undergo deformation and/or fracture processes depending on their mechanical properties [13].

Despite of the fabrication method used, low wettability between Al/C is a serious problem; it depends on the particle characteristics like: type, shape, size, surface roughness and surface chemistry [4]. In order to modify wettability in this system, some methods have been used; one of these is the incorporation of a metal with aluminum affinity to form metal-coated graphite particles [9]. Nickel coated graphite particles have been produced and tested by Ip et al. [7]. In the present work, the effect of milling time and metal particle concentration on the microstructure of graphite-metal couples produced by a solid state route (MM) is investigated. Characterization of the couples produced is performed using scanning electron microscopy (SEM), energy dispersive X-ray spectrometry and X-ray diffraction.

EXPERIMENTAL PROCEDURE

Pure graphite powder (99.9% purity, -20 +84 mesh) and gas atomized metal powders Cu (99.5% purity, -150 mesh), Ag (99.9% purity, -325 mesh) and Ni (99.8% purity, -200mesh) from Alfa Aesar were used in the present investigation. Sample mixtures with low and high metal concentrations of 10 and 15 at.% are prepared by mixing previously weighted graphite and metal powders according with Table I. The powder mixtures are then mechanically milled using a high-energy SPEX 8000M mill, under an inert argon atmosphere and a ball to powder ratio of 5:1 (in weight). Hardened steel vial and Fe-Cr balls are utilized as milling media. Four milling times are used (0, 1, 4 and 8h). Pure graphite powder samples are prepared as reference material for comparison proposes.

The morphology, size and metal-particle distribution of milled powder mixtures is examined using a JEOL-JSM 7201F SEM equipped with an EDS microanalysis system. For SEM examination a small amount of the milled powder is placed on carbon tapes. X-ray diffraction analysis is carried out using a PANalytical X´pert PRO X-ray diffractometer with CuKα radiation ($\lambda = 0.15406$ nm) operated at 40 kV / 35mA and at a scanning speed of 50 s/ step for a scanning range of 20–100° in steps of 0.0167°.

The powder mixtures are then mechanically milled using a high-energy SPEX 8000M mill, under an inert argon atmosphere, with a ball to powder ratio of 5:1 (in weight). Hardened steel vial and Fe-Cr balls are utilized as milling media. Four milling intervals are used (0, 1, 4 and 8h). Pure graphite samples (without metal addition) are prepared as reference material for comparison proposes. The morphology, size and metal-particle distribution of milled powder is examined using a JEOL-JSM 7201F scanning electron microscope (SEM) equipped with an energy dispersive spectrometer (EDS). For SEM observation, a small amount of the milled powder is placed on carbon tapes. Cross section samples are obtained by mounting the milled powders in a conductive thermoplastic and prepared by standard metallographic procedures.

Table I. Graphite-Metal couples composition and nomenclature (where H = High metal concentration, L = Low, C = pure graphite, N = Ni, C = Cu, S = Ag and x = 0, 1, 4, 8h).

Sample	Additive	Metal Conc. [at.%. – wt.%]
Cx	n.a.	n.a.
HNx	Ni	15.0 - 46.31
LNx		10.0 - 35.19
HCx	Cu	15.0 – 48.28
LCx		10.0 – 37.02
HSx	Ag	15.0 - 61.32
LSx		10.0 - 49.95

RESULTS AND DISCUSSION

Morphological analysis. Metal powders are spherical in shape and exhibit particle sizes of 100, 50 and ~5 μm for Cu, Ni and Ag, respectively (see Fig. 1). Graphite powders are irregular in shape and have a particle size >100μm (Fig. 2).

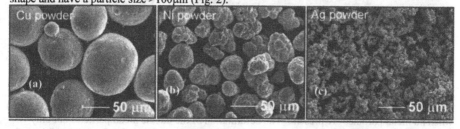

Fig. 1. SEM micrographs showing the morphology of as-received metal powders: (a) Cu,)b) Ni, (c) Ag.

Figure 2 shows the effect of milling time on the morphology of graphite particles. The particles are fragmented during milling due to their brittle behavior and, as a result, the particle size decreases continuously with milling time. The presence of metal particles mixed with graphite particles changes the trend observed in Fig. 2. During the initial stage of milling of the metal-graphite mixtures, the ductile metal particles undergo deformation while the brittle graphite particles undergo fragmentation. The ductile metal powder particle size changes as a result of two opposing effects: cold welding increases while fracturing reduces the particle size.

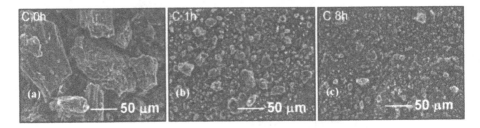

Fig. 2. Effect of milling time on the morphology of graphite powders: (a) as-received, (b) 1 hour, (c) 8 hours.

During the early stages of milling the powder metal particles are ductile and cold welding predominates producing large laminar particles which increase the overall particle size of the mixture (Fig 3a).

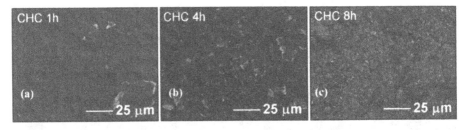

Fig. 3. Effect of milling time on the morphology of a graphite-15%at. Cu powder mixture (SEM backscattered micrographs): (a) 1 hour, (b) 4 hours, (c) 8 hours.

With increasing milling times the high proportion of graphite particles and the work hardening of the metal particles increase the rate of fragmentation by particle fracturing which then leads to a reduction in particle size [15]. Comparing Figs. 2c and 3c shows that adding 15% Cu particles to the graphite powder produces a finer powder mixture after 8 hours of milling than when no metal particles are present. The deformation of the ductile metal particles not only consumes part of the milling energy but also increases the magnitude of deformation of the softer particles and, therefore, accelerates the global process and leads to a smaller final particle size for the same milling conditions.

As the milling process proceeds, a balance is established between cold welding and fracturing events and a steady-state regime is reached. The particle size is stabilized and no further change is observed with increasing milling time. The stable particle size is usually known as the comminution limit [15]. As shown in Fig. 3c the distribution of bright Cu particles in the graphite matrix (grey) is very uniform after 8 h of milling. In contrast, for shorter milling times (Figs. 3a and 3b) metal particles are not uniformly distributed and can easily be distinguished from graphite particles.). As shown in Fig. 3b, milling during 4 hours produced a change in particle morphology and the metal phase cannot easily be distinguished from the graphite phase; the larger particle flakes exhibit a range contrast levels indicating possibly the occurrence of a diffusion process during milling. Repeated fracturing of these composite particles and convolution results in a uniform distribution of the particles as observed in Fig. 3c for longer milling times.

Figure 4 compares the microstructures of C-15%at Ag and C-15%at Ni couples after milling during 8 hours. As can be seen, the samples exhibit different particle size and metal particle distribution. The Ni particles are larger and not as uniformly distributed as the Ag particles in the matrix. The laminar morphology of the Ni particles in Fig. 4b indicates that plastic deformation predominates and that the C-Ni sample is at an earlier stage the milling process.

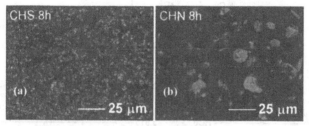

Fig. 4. Morphology of a) graphite-15%at Ag and b) graphite-15%at Ni powder mixtures after 8 hours of milling (SEM backscattered electron micrographs).

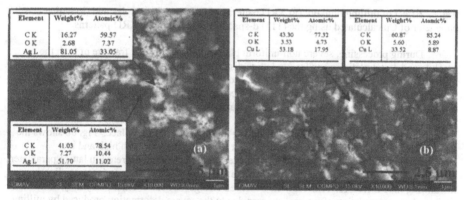

Fig. 5. SEM backscattered electron images and EDS spectra of selected zones in cross sections of samples of a) C-15%at Ag and b) C-15%at Cu produced by milling during 8 hours

Figure 5 shows SEM backscattered electron images and EDS spectra obtained on selected zones of the C-15%at Cu and C-15%at Ag samples produced by milling during 8h. The presence of both C and Ag and C and Cu in the analyzed regions of each sample (bright and gray zones) suggests that the milling process is effective in producing a graphite-metal composite material. However, the presence of oxygen in the EDS-microanalyses clearly suggests that some oxidation of the metal particles has taken place. Also, the absence of Fe and/or Cr in the microanalyses also indicates that no significant contamination (by wear and tear of milling media) of the milled powders has occurred.

X-Ray Diffraction (XRD). Figure 6a illustrates the effect of milling time on the XRD patterns of C-15%at Cu. A similar behavior was observed for C-Ni and C-Ag powder mixtures. As can be seen, all the graphite XRD peaks decrease in intensity while those corresponding to the Cu phase change very little, a small increase is observed at the beginning of milling.

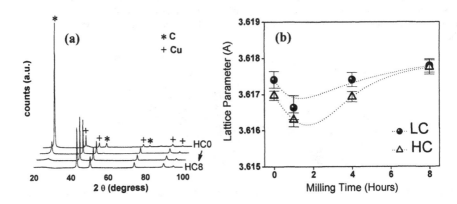

Fig. 6. Effect of milling time on: (a) XRD patterns of C-15%at Cu couples and (b) lattice parameter of Cu calculated using the (111) d-spacing observed in the XRD patterns.

During processing, severe non-uniform plastic deformation and fracture of both the graphite and metal particles can lead to lattice distortion, particle size refining and accumulation of internal stress that may result in XRD peak broadening and shift and reduced maximum intensities. Peak shift indicates a change in lattice parameter [15]. A change in lattice parameter may also result of formation of solid solutions during milling. No evident shift in the position of the Cu peaks is observed with increasing milling time. This observation suggests that incorporation of C atoms to the Cu lattice during milling does not take place. Fig. 6b shows the variation of the Cu lattice parameter with respect to milling time for both high and low concentrations of this element in the C-Cu powder mixture. A small decrease in lattice parameter is observed after 1h of milling and, after that, the lattice parameter seems to increase slightly as the milling time increases. Although the observed changes are rather small, they may be attributed to accumulation of internal stresses due to the severe deformation produced by milling. . Several factors influence the detection limit of a given phase by XRD. The effect of milling time on the graphite XRD peaks observed in Fig. 6a may be due to broadening effects due to the severe reduction of the crystallite size in this component.

CONCLUSIONS

Graphite–Metal couples are produced using mechanical milling as a preparation method. Metal particles in the prepared graphite-metal samples exhibit sub-micron sizes and are uniformly distributed in the processed powders. The results show that metal particle concentration and milling time have important effects on the refinement behavior of samples.

Increasing concentration of metallic particles accelerates the milling process leading to faster particle fracture and uniform metal distribution at longer milling times. The milling time needed to reach this condition appears to depend on the type of metal particles added to the graphite powder.

ACKNOWLEDGMENTS

This research was supported by CONACYT (106650). Thanks to D. Lardizabal-Gutierrez, W. Antúnez-Flores, K. Campos-Venegas and E. Torres-Moye for their valuable technical assistance.

REFERENCES

1. K. Kondoh, H. Fukuda, J. Umeda, H. Imai, B. Fugetsu, M. Endo. Mat. Sci. Eng. A, **527**, 4103 –4108 (2010).
2. Q. Li, C. A. Rottmair, R. F. Singer. Comp. Sci. Tech. (2010). In Press.
3. K.R. Ravi, V.M. Sreekumar, R.M. Pillai, C. Mahato, K.R. Amaranathan, R. Arul Kumar, B C. Pai. Mat. Design, **28**, 871–881 (2007).
4. O. Yılmaz, S. Buytoz, Com. Sci. Tech. **61**, 2381–2392 (2001).
5. M.V. Achutha, B.K. Sridhara, D. Abdul Budan, Mat. Design **29**, 769–774 (2008).
6. J.M. Torralba, C.E. da Costa, F. Velasco. J. Mat. Proc. Tech. **133**, 203–206 (2003).
7. S.W. Ip, R. Sridhar, J.M. Toguri, T.F. Stephenson, A.E.M. Warner. Mat. Sci. Eng. A **244**, 31–38 (1998).
8. H. Mayer, M. Papakyriacou. Carbon **44**, 1801–1807 (2006).
9. F. Akhlaghi, S.A. Pelaseyyed. Mat. Sci. Eng. A **385**, 258–266 (2004).
10. T.G. Durai, Karabi Das, Siddhartha Das. Mat. Sci. Eng. A **445–446**, 100–105 (2007).
11. A. Rodríguez-Guerrero, S.A. Sánchez, J. Narciso, E. Louis, F. Rodríguez-Reinoso. A. Mater. **54**, 1821–1831 (2006).
12. S.P. Sharma, S.C. Lakkad. Surface & Coatings Tech. (2010). In Press.
13. J.B. Fogagnolo, E.M. Ruiz-Navas, M.H. Robert, J.M. Torralba. S. Mater. **47**, 243–248 (2002).
14. A. Daoud. Mat. Lett. **58**, 3206–3213 (2004).
15. C. Suryanarayana. Prog. Mat. Sci. **46**, 1-184 (2001).
16. A.M.K. Esawi, K. Morsi, A. Sayed, M. Taher, S. Lanka. Comp. Sci. Tech. (2010). In Press

Mater. Res. Soc. Symp. Proc. Vol. 1276 © 2010 Materials Research Society

Strengthening phases in the production of Al$_{2024}$-CNTs composites by a milling process.

R. Pérez-Bustamante[1], F. Pérez-Bustamante[2], J. M. Herrera-Ramírez[1], I. Estrada-Guel[1], P. Amézaga-Madrid[1], M. Miki-Yoshida[1], R. Martínez-Sánchez[1]

[1]Centro de Investigación en Materiales Avanzados (CIMAV), Laboratorio Nacional de Nanotecnología, Miguel de Cervantes No. 120, C.P. 31109, Chihuahua, Chih., México.
[2]Instituto Tecnológico de Chihuahua (ITCH), Av. Tecnológico No. 2909, C.P. 31310, Chihuahua, Chih., México. e-mail: roberto.martinez@cimav.edu.mx

ABSTRACT

Carbon nanotubes (CNTs) synthesized by a chemical vapor deposition (CVD) method and the 2024 aluminum alloy (Al$_{2024}$) are used in the production of Al$_{2024}$-CNTs composites. An homogeneous dispersion of the CNTs into the aluminum matrix is achieved by a mechanical milling processing. CNTs keet their morphology after milling and sintering processes. Formation of aluminum carbide as a function of CNTs contents is observed. Formation of equilibrium phases during sintering is observed by electron microscopy. CNTs and aluminum carbide in the composites are characterized by transmission electron microscopy. Hardness results of sintered products show an increment of up to 285% over the unreinforced alloy prepared by the same route.

INTRODUCTION

The wide use of aluminum alloys in several industrial fields has attracted the attention of a great number of researchers focused in the improvement of their mechanical properties [1-5]. The 2XXX and 7XXX series are of special interest because their mechanical enhancement lies in the precipitation of phases into the aluminum matrix during heat treatments. Additionally to the conventional routes used to fabricate these alloys, mechanical alloying (MA) offers the possibility for their production. This technique is widely used for the dispersion of strengthening phases, like nanoparticles or nanofibers, whose function is to get an increment in the mechanical behavior of an alloying system [6, 7]. In this regard, CNTs have emerged as a promising reinforcement media since their discovery in 1991 for the production of metal matrix composites. Nevertheless, research on this topic has received only modest attention. One of the main reasons is the lack of a good dispersion technique of CNTs into the metal matrix by conventional metallurgy. However, this problem can be solved by the use of MA. Works carried out about these topics demonstrate that the composites mechanical performance improve when the CNTs dispersion is achieved by milling processes [8-13].

EXPERIMENTAL PROCEDURE

CNTs prepared by chemical vapor deposition (CVD) are used as reinforcement material in order to produce Al$_{2024}$-based nanocomposites. Table I shows the composition of the Al$_{2024}$ matrix prepared from pure elemental powders. Different CNTs concentrations are studied: 0.0, 0.5, 1.0, 2.0, 3.0, 4.0 and 5.0 wt. %. The milling time is varied to achieve the CNTs dispersion to

5, 10, 20 and 30 h. Table II shows the identification code for alloys and composites in this investigation. Each mixture is mechanically milled in a high-energy mill (SPEX-8000M).

Table I. Composition of the 2024 aluminum alloy (wt. %)

Al	Cu	Mg	Mn	Ti	Zn
93	4.5	1.5	0.6	0.15	0.25

Table II. Content (wt. %) and milling time (h) used for MA of Al$_{2024}$-CNTs composites and alloys

CNTs	Milling time (h)			
(wt. %)	5	10	20	30
0.0	A$_{00}$	B$_{00}$	C$_{00}$	D$_{00}$
0.5	A$_{05}$	B$_{05}$	C$_{05}$	D$_{05}$
1.0	A$_{10}$	B$_{10}$	C$_{10}$	D$_{10}$
2.0	A$_{20}$	B$_{20}$	C$_{20}$	D$_{20}$
3.0	A$_{30}$	B$_{30}$	C$_{30}$	D$_{30}$
4.0	A$_{40}$	B$_{40}$	C$_{40}$	D$_{40}$
5.0	A$_{50}$	B$_{50}$	C$_{50}$	D$_{50}$

To avoid oxidation of the powder mixture during the milling process, a static argon atmosphere is used in all runs. The apparatus and milling media are made of hardened steel. The milling ball-to-powder weight ratio is set at 5:1. In order to avoid excessive welding of particles, methanol is added to the powders to act as a process control agent. The as-milled products are cold compacted under uniaxial load pressure of ~3 tons during one minute. Discs of 5 mm in diameter and ~1 mm in height are obtained and then sintered during 2 h at 773 K under argon atmosphere. Hardness of the sintered products is determined with a microdurometer Future-Tech model FM-7 using 200 g of load and 15 s of dwell time. The average values of at least five points of randomly selected regions in each sample are reported. Microstructural observations are performed by transmission electron microscopy (TEM) in a JEOL JEM 2200FS operated at 200 kV. For TEM observations, samples are prepared by focused ion beam (FIB) in a JEOL JEM 9320-FIB operated at 30 kV for thick cuts and 5 kV for final polishing at 25 A. X-ray diffraction (XRD) is carried out in a Siemens D5000 diffractometer with Cu Kα radiation (λ=1.5406 Å) and operated at 40 kV and 25 mA in the 2θ range of 20-80°. The step and acquisition time are 0.05° and 5 s, respectively.

RESULTS AND DISCUSSION

XRD patterns corresponding to the milled powders and sintered products, as a function of milling time and CNTs concentration, are displayed in Fig. 1. For comparison purposes, the sample NM is prepared. This corresponds to a powder mix with the Al$_{2024}$ composition but without CNTs. Characteristic reflections of Al, Cu and Mg are observed for the NM alloy after milling for 5 h (Fig. 1a). However after 30 h of milling, there is formation of a solid solution (Fig. 1b). For sintered products (Figs. 1c and 1d), Al$_2$Cu phase precipitation is observed, which takes place during the slow cooling of the sintering process. No carbon reflections are observed from X-ray patterns for any sample. However, the formation of the Al$_4$C$_3$ phase suggests the presence of a reaction between CNTs and aluminum. The intensity of the aluminum carbide signal increases as the CNTs content increases.

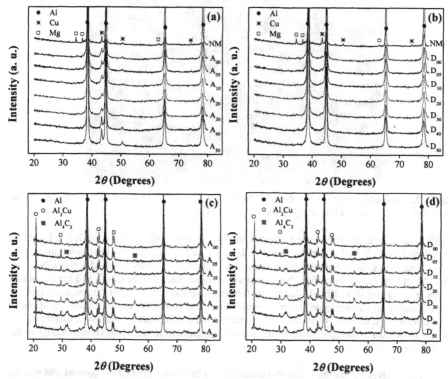

Figure 1. XRD spectra from the Al_{2024}-CNTs composites. (a) 5 h and (b) 30 h of milling in the as-milled condition. (c) 5 h and (d) 30 h of milling in the sintered stage.

Microstructural observations are displayed in Fig. 2 for the A_{50} and D_{50} composites in both the as-milled and sintered conditions. For milled powders, a quasi-lamellar structure is presented in the A_{50} composite, which is produced by the repeated cycle of fracture-welding occurred during the early stages of the MA process [14]. The D_{50} composite presents some remnant Mn particles well distributed into the aluminum matrix. This indicates that even though a solid solution is observed from XRD spectra, a prolonged milling time is necessary to achieve a complete dissolution of the alloying elements. The analysis of the sintered products (A_{50} and D_{50}) shows the presence of precipitated phases (Figs. 2c and 2d), which are mainly constituted by Mn (marked with circles) and Cu (marked with a square). No evidence of the aluminum carbide is observed in SEM analysis. No significant microstructural changes are observed in the sintered composites even though a variation in the products density is displayed.

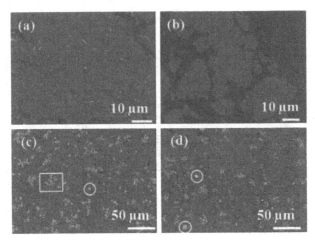

Figure 2. Secondary electron SEM micrographs. (a) A_{50} and (b) D_{50} samples in the as-milled condition. (c) A_{50} and (d) D_{50} composites in the sintering stage.

Bright field TEM micrographs are presented for the A_{50} composite in the as-milled and sintered conditions in Fig. 3. Isolated nanotubes are observed well embedded into the aluminum matrix. This corroborates their homogeneous dispersion achieved by MA. Well defined CNTs are observed in Fig. 3a, while in Fig. 3b a stronger interaction between the CNTs and the aluminum matrix is observed; this effect is due to the further integration of the CNTs outer walls into the metal, forming as consequence stronger bonds than those formed in the as-milled condition. In Fig. 3c, aluminum carbide formation is observed shaped as long bars of ~15 nm in diameter. The interval between aluminum carbide lattice fringes is measured as 0.83 nm, corresponding to the (003) plane.

Fig. 4 gives the hardness results presented as a function of the CNTs content and milling time. A rapid increment in the hardness values is seen as the CNTs content increases. The increment in the milling time seems to have no considerable effect on the products hardness behavior. For unreinforced alloys, the highest hardness value (85.9) is reached by the A_{00} alloy and the lowest value (67.1) by the B_{00} alloy. The maximum hardness is achieved by the B_{50} composite with 290.9 units. For the same composition of CNTs, D_{50} composite presents 280 hardness units. The strongest composite (B_{50}) represents an increase of 285% over the unreinforced alloy milled for the same time (B_{00}). An important aspect is that the composites with CNTs higher than 2 wt. % present similar or higher Vickers hardness values than those of the Al_{2024}-T86 and Al_{2024}-T6, which are among the strongest commercial aluminum alloys. In addition, it is well known that Al_{2024} alloys are susceptible to precipitation strengthening; thus, we expect an additional increment in hardness values by heat treatment.

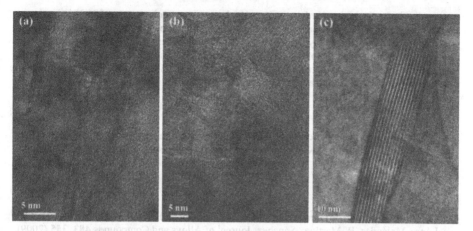

Figure 3. Bright field TEM micrographs of the A_{50} composite in the (a) as-milled condition and (b) sintered condition. (c) Aluminum carbide into the A_{50} composite after sintering process.

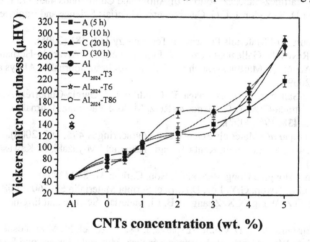

Figure 4. Vickers microhardness as a function of the CNTs content and milling time.

CONCLUSIONS

Al_{2024}-CNTs composites have been produced successfully by mechanical alloying. SEM analysis shows a quasi-lamellar structure in composites milled for 5 h. For 30 h of milling some remaining Mn particles are observed. In the sintered condition, the low cooling in the sintering process lead to the precipitation of phases constituted mainly by Cu and Mn. TEM observations show a homogeneous dispersion of CNTs. In addition, XRD patterns present characteristic reflections attributed to the Al_4C_3, whose intensity grows as a function of the CNTs content. B_{50}

composite shows the higher hardness value, which represents an increase of 285 % in hardness over a C phases-free alloy produced by the same route and milled for the same time. Thus the mechanical behavior reached by the Al_{2024}-CNTs composites here investigated is due mainly to the homogeneous distribution of the CNTs, promoted by MA and the formation and dispersion of aluminum carbide in the metallic matrix. Aluminum carbide is most likely formed through the interaction between aluminum and carbon nanotubes during the sintering process.

ACKNOWLEDGEMENTS

This research is supported by CONACYT (106658). Thanks to W. Antúnez-Flores, E. Torres-Moye, O. Solis-Canto, K. Campos-Venegas and C. Ornelas-Gutierrez for their valuable technical assistance.

REFERENCES

1. C. Carreño-Gallardo, I. Estrada-Guel, M.A. Neri, E. Rocha-Rangel, M. Romero-Romo, C. López-Meléndez, R. Martínez-Sánchez, Journal of Alloys and Compounds **483**, 355 (2009).
2. I. Estrada-Guel, C. Carreño-Gallardo, D.C. Mendoza-Ruiz, M. Miki-Yoshida, E. Rocha-Rangel, R. Martínez-Sánchez, Journal of Alloys and Compounds **483**, 173 (2009).
3. J. Oñoro, M.D. Salvador, L.E.G. Cambronero, Materials Science and Engineering A **499**, 421 (2009).
4. M. Kok, Journal of Materials Processing Technology **161**, 381 (2005).
5. R. Goytia-Reyes, V. Gallegos-Orozco, H. Flores-Zuñiga, F. Alvarado-Hernandez, R. Huirache-Acuña, R. Martínez-Sánchez, A. Santos-Beltrán, Journal of Alloys and Compounds **485**, 837 (2009).
6. R. Martínez-Sánchez, J. Reyes-Gasga, R. Caudillo, D.I. García-Gutierrez, A. Márquez-Lucero, I. Estrada-Guel, D.C. Mendoza-Ruiz, M. José Yacaman, Journal of Alloys and Compounds **438**, 195 (2007).
7. Hisao Uozumi et al., Materials Science and Engineering A **495**, 282 (2008).
8. Hansang Kwon, Mehdi Estili, Kenta Takagi, Takamichi Miyazaki, A. Kawasaki, Carbon **47**, 570 (2009).
9. Rong Zhong, Hongtao Cong, Pengxiang Hou, Carbon **41**, 848 (2003).
10. H.J. Choi, G.B. Kwon, G.Y. Lee, D.H. Bae, Scripta Materialia **59**, 360 (2008).
11. C.F. Deng, D.Z. Wang, X.X. Zhang, A.B. Li, Materials Science and Engineering A **444**, 138 (2007).
12. R. Pérez-Bustamante, C.D. Gómez-Esparza, I. Estrada-Guel, M. Miki-Yoshida, L. Licea-Jiménez, S.A. Pérez-García, R. Martínez-Sánchez, Materials Science and Engineering A **502**, 159 (2009).
13. R. Pérez-Bustamante, I. Estrada-Guel, W. Antúnez-Flores, M. Miki-Yoshida, P.J. Ferreira, R. Martínez-Sánchez, Journal of Alloys and Compounds **450**, 323 (2008).
14. C. Suryanarayana, Progress in Materials Science **46**, 1 (2001).

Mater. Res. Soc. Symp. Proc. Vol. 1276 © 2010 Materials Research Society

Wear Properties of an In-Situ Processed TiC-Reinforced Bronze

R. Sanchez and H. F. Lopez

Materials Department, University of Wisconsin-Milwaukee, Milwaukee WI 53209.

ABSTRACT

In this work, an Al-bronze alloy is reinforced with TiC through reaction of the alloy melt with methane gas. The resultant alloy is then centrifugally cast in cylindrical molds. It is found that the surface at the inner diameter of the cast contained in-situ produced TiC as well as Fe-rich inclusions. Metallographic observations using optical and scanning electron microscopy confirmed the presence of TiC particles (30 % volume), alpha and beta grains including iron precipitates. Cylindrical pins are machined from the inner surface and tested under various conditions in a three pin on disk Falex machine. Pins are tested under a constant load of 2.86 MPa and friction and wear rates are determined from measurements of weight losses versus wear lengths. It is found that under the applied load the reinforced material exhibits high friction and relatively low wear when compared with the unreinforced material. Apparently, under these conditions the TiC particles become abrasive particles thus contributing to wear of the steel counter-face through three body abrasive wear.

INTRODUCTION

Aluminum bronzes are copper-base alloys in which up to 14% Al is added as the main alloying element. Smaller additions of nickel, iron manganese and silicon are often introduced to meet different requirements of strength, ductility and corrosion resistance [1,3]. In particular, they exhibit good corrosion resistance in sea water and relatively high strengths. In these alloys, Al has a marked effect on the corrosion properties through the formation of an aluminum oxide layer which seals the core alloy from any further corrosive reactions [4].

However, the wear resistance of conventional Al-bronzes is not always the one required for marine applications such as anti-slake devices and hauling winches [5]. Asbestos and lead are the typical materials used for these types of devices. Nevertheless, these materials are currently unacceptable due to their toxicity and damage to the environment. Hence, TiC reinforced Al-bronzes have been considered as an attractive alternative for the there manufacture of these components. Among the manufacturing methods considered is the Reactive Gas Injection method (RGI [5].

The RGI method is based on the "in-situ" processing of carbide reinforced Cu alloys [6]. In this process, CH_4 is injected into the Al-Cu-melt where an "in-situ" reaction with a purposely added solute (Ti) gives rise to a distribution of TiC particles within the bronze matrix. Thermodynamically, the process is highly stable and the interfaces developed between matrix and reinforcement are tightly bonded. This is in contrast with other conventional processes where, the reinforcement particles are introduced by external means. The objective of the present work is to investigate the wear and friction resistance of an "in-situ" TiC-Al-bronze composite processed by centrifugal casting and of the as-cast Al-bronze under similar wear conditions.

EXPERIMENTAL

Aluminum Bronze pin samples (as-cast aluminum bronze and reinforced aluminum bronze with a 3.1 wt% TiC) are machined to 6 mm diameter approximately and 20 mm length from the test castings. The as-cast aluminum bronze and reinforced aluminum bronze pins are tested separately. Three pins are fixed into the wear testing fixture, and they are ground and polished all the way to 0.5 μm alumina suspension. Measurements of the surface roughness using a Mitutoyo Surftest Analyzer, series 178 are between 0.025 to 0.05 μm (1 to 2 μin), in agreement with the ASTM standard 732-82. The counter disc material is quench hardened 4340 steel (with a Rockwell C hardness of 52). The dimensions of the disk are 50 mm diameter and 5 to 10 mm thick. A three pin-on-disc FALEX machine is used for friction and wear measurements. In this work, the contact stress is kept constant at is 2.86 MPa.

Metallographic sections of non-reinforced and reinforced cast samples are metallographically prepared for observations in an optical and a scanning electron microscope (SEM). The microstructure is resolved by polished and then etching with potassium dichromate + HCl. The present phases are examined in an SEM Topcon ABT 32 and by EDS means.

RESULTS AND DISCUSSION

The chemical composition of the base Al-Bronze composite is given in Table I. Notice the relatively high amounts of Fe in this alloy. The dispersion of reinforcement particles through the matrix is largely uniform; however there are areas of agglomeration in the sample (Figs. 1a-b).

Table I. Chemical composition of the Aluminum Bronzes.

	Bronze	UNS C95400
Aluminum	10.7	11.0 Max.
Copper	83.5	85.0 Max.
Iron	4.48	4.0 Max.

The TiC particle sizes in this composite range from 0.25 to 40 μm, with the main variation in particle size due to particle clustering as a result of the centrifugal casting process (see Fig. 2).

Wear behavior

The exhibited friction coefficient values for the as cast Al-Bronze and TiC reinforced Al-Bronze are shown in Fig. 3. Notice that the frictional resistance of the as-cast aluminum bronze is below the one exhibited by the reinforced Al-Bronze. Prior to the surfaces beginning to move relative to each other, small contact areas between the mating surfaces become fused together (junctions). Thus, when the machine applies a force to break these junctions, the resulting stresses in the metal are small, only small metal fragments become detached. In the case of the as-cast Al-Bronze, these fragments or particles are quickly transferred from the softer metal (Al-Bronze) to the harder metal (steel). They adhere firmly to the steel in the form of a thin layer and are work-hardened.

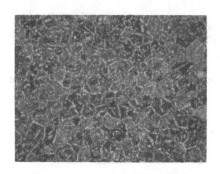 

(a)   (b)

Figure1. Low magnification of etched microstructures. (a) Unreinforced and (b) reinforced Al-Bronze. Ethant:.Potassium Dichromate (100X).

Figure 2. Optical micrograph showing TiC agglomerates in the reinforced Al-Bronze, (200X).

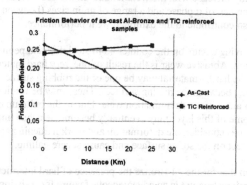

Figure 3. Friction coefficient values of as cast Al- Bronze and TiC reinforced Al-Bronze rubbing against 4340 steel (R_c 52) under 2.86 MPa, at 300mm/sec sliding velocity in dry air.

Hence, the newly transferred particles agglomerate within the existing transferred layer. Some particles may transfer back to the aluminum bronze. The weight losses observed for the as-cast aluminum bronze are much higher than those of the Al-Bronze composite (see Fig. 3). It is expected that a high amount of α (over 70%), which is Cu-rich phase, renders the structure soft, and more prone to adhesion. In contrast, a high proportion of β (rich-Al phase) and a small grain size would render the structure hard and brittle.

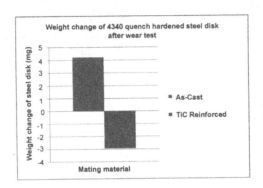

Figure 4. Weight changes in the 4340 steel counter disk after sliding 25 km under 2.86 MPa constant stress, at 300mm/sec sliding velocity in dry air.

Wear performance is not related to matrix hardness alone but relies on the combination of hard Fe_3Al (κ particles) in this alloy, embedded in a relatively soft α matrix. Figure 5 shows the wear volume losses for the as-cast and TiC-reinforced Al-Bronze. Notice that the reinforced Al-bronze exhibits relatively low wear losses for lengths of up to 25 km. In contrast, the wear volume losses continually increase with the wear length. In the TiC reinforced Al-Bronze a combination of a volume pct. of α phase and a range of grain sizes (between 33 and 46 μm) including κ particles also contribute to the exhibited wear performance.

In the present case, grooving occurs on the counter 4340 steel disc and peeling of TiC particles can cause severe abrasion. Abrasive wear is the result of one very hard material plowing grooves into a softer material. The harder material may be one of the rubbing surfaces or hard particles that have found their way between the mating surfaces. These may be foreign particles or particles resulting from adhesive or delamination wear. Due to the build up of elastic energy in the transferred layer, some of this layer may eventually become detached and form small debris. This debris then undergoes considerable deformation and work hardening and are therefore liable to have and abrasive effect on the softer surface and cause severe galling.

It is already been shown that for the Al-Bronze the presence of hard intermetallic particles in the soft constituent of the microstructure is an advantageous feature for resisting wear [4]. The current samples show mixed results. Initial friction coefficients for the TiC reinforced material is slightly lower than for the as-cast condition. However, as testing continues, the friction coefficient for the TiC samples remains fairly steady while the as-cast friction coefficient decreases significantly (see Fig. 3).

The as-cast pins showed fractured layers of bronze which are smeared on the counter steel disk. The metal losses are very high in this case but no grooves are noticed. The aluminum bronze composite pins did not show any grooves or any kind of adhesive wear, only a few scratches are seen on the surface. This is a large difference that illustrates the significant impact of adding TiC reinforcements. In the case of the Al-bronze-TiC composite it is found that it posses an inherent frictional resistance to plowing.

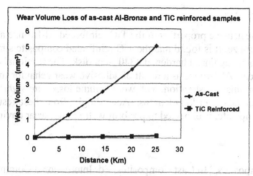

Figure 5. Weight changes of 4340 quench hardened steel (R_c 52) after wearing against as cast Aluminum Bronze and TiC reinforced Aluminum Bronze composite under 2.86 MPa constant stress, at 300mm/sec sliding velocity with a final traverse distance of 25 Km in dry air.

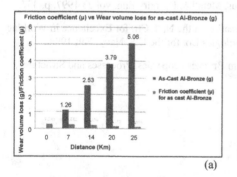 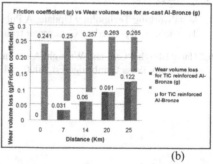

(a)   (b)

Figure 6. Friction coefficient values compared to Wear volume losses for (a) As-cast Al-Bronze and (b) TiC reinforced Aluminum Bronze composite rubbing against 4340 quench hardened steel (R_c 52) under 2.86 MPa constant stress, at 300mm/sec sliding velocity in dry air.

Figure 6 shows how the friction coefficient relates to the wear volume losses for the as-cast Aluminum Bronze and the reinforced TiC Aluminum Bronze. In the reinforced TiC Al-Bronze a trend is observed as the friction coefficient increases the wear volume losses increases due to the abrasive wear experienced by the samples. In contrast, the as-cast Al-Bronze does not show the same trend because as the volume loss increases considerably the friction coefficient is decreasing due to the adhesive wear the samples experience in contact with the 4340 steel disk. Based on this behavior, hardness of the samples has a considerable effect on the wear properties. TiC particles increase the hardness of the metal matrix while in the as-cast samples containing a soft metal matrix rubbing against a harder counter surface leads to adhesive wear.

CONCLUSIONS

It is found that the wear resistance properties of the TiC reinforced Al-Bronze are superior to those of the as-cast Al-Bronze. It is found that the TiC reinforced composite surface did not scratch deeply when wearing against a hardened 4340 steel disk. The frictional coefficient is relatively low for the as-cast Al-Bronze as a result of adhesive wear behavior for the contact stress and speed chosen in this investigation. Yet, wear volume losses are significantly high in the as-cast material particularly over long lengths. Accordingly, from these results, the TiC reinforced Al-Bronze composite, is expected to perform well in high wear environments.

REFERENCES

1. E. Fras, "In Situ Composites," In Cast Composites, Ed. International Committee of Foundry Technical Association, CIATIF, 1995, 23-38.

2. S. Sarker and A. P. Bates, British Foundryman 1967, vol. 60, p.30.

3. R. Thomson, Modern Castings, 1968, vol. 53, 189-199.

4. E. Fras, A. Janas, H. F. Lopez and A. Kolbus, Metall & Foundry Eng, vol.23 1997, p. 357.

5. Metalworking Technology Update, A Publication of the Natl. Ctr. for Excellence in Working Technology. Operated by Concurrent technologies Co. for the U.S. Navy, Fall 1998.

6. Harry Meigh, "Cast and Wrought Aluminum Bronzes Properties, Processes and Structure", University Press Cambridge Ed. 2000, p.4.

Mater. Res. Soc. Symp. Proc. Vol. 1276 © 2010 Materials Research Society

Nanoceria Coatings and Their Role on the High Temperature Stability of 316L Stainless Steels.

H. Mendoza-Del-Angel and H. F. Lopez.

Materials Department, University of Wisconsin-Milwaukee, 3200 N. Cramer St. Milwaukee WI 53211

ABSTRACT

In this work, a method is considered to produce uniform nanoceria surface coatings on 316L stainless steel such as dipping. Coated steels using the aforementioned method are exposed to high temperature 800-1000°C and their oxidation behavior is investigated. It is found that the nanoceria particles in the implemented coatings exhibit some growth during high temperature exposure. In addition, thermogravimetric determinations of oxidation resistance in coated and bare samples at 900°C clearly indicates that the nanoceria coated stainless steels exhibits a two fold reduction in mass gain when compared with bare ones. Optical and scanning electron microscopy are employed to characterize the developed oxide scale morphologies. It is found that in areas are nanoceria is not uniformly coated, Fe-rich oxide islands develop, whereas in coated regions the scale is Cr and Ce rich indicating that the scale is probably a Ce doped Cr oxide.

INTRODUCTION

Structural steels for high temperature applications are prone to undergo oxidation and corrosion in severe environments. The nature of the reaction products is complex and the rate at which the materials corrode vary widely. Therefore, it is highly desirable to improve high temperature alloy properties, as well as the capacity for prediction oxidation/corrosion rates.

In particular, corrosion has been a serious problem with significant impact in the operation of conventional power generation plants. The high Temperatures of the process and the complexity of the corrosive environment make it difficult to design for long term operations using conventional steels.

Other applications where improved performance is sought are in the field of fuel cells. These cells are units used to generate electricity through an electrochemical process. They do not have any combustion and they produce low emissions, but they operate at high temperatures (659 to 1000°C). These high temperatures are a potential challenge for the metallic alloys (austenitic and ferritic stainless steels) to stand corrosion in the interconnectors and heat exchangers.

Stainless steels are the materials of choice for high temperature applications as they can be used in a wide variety of environments. They are fairly inexpensive and posses good oxidation/corrosion resistance. However, it is desirable to enhance the long term oxidation resistance of these steels. A common approach has been the use of surface coatings [1, 2] and it is the approach employed in this work. Accordingly, in the present work nanoceria coatings are employed to protect a 316 stainless steel from high temperature oxidation in dry air. In particular, the work is focused on identifying conditions that will give rise to a uniform nanoceria surface coating and the effect of the coating on improving the oxidation resistance of the 316 stainless steel.

EXPERIMENTAL

Test coupons are cut from the 316 stainless steel sheets and then grinded using 600 and 1200 grit SiC paper followed by ultrasonic cleaning in isopropanol. Nanocrystalline CeO_2 particles are synthesized using the micro-emulsion method published by Patil *et al.* [3] and Wu *et al.* [4]. A solution containing the nanocrystalline CeO_2 is extracted from the upper layer of the micro-emulsion system and collected as a coating solution.

The nano-particle sizes are dispersed by ultrasound and measured using a light scattering technique based on the Malvern Instrument Zetasizer Nano ZS and the results are shown in Fig. 1.

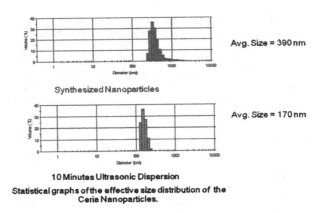

Figure 1. Nanoceria size distributions (a) after processing and (b) after ultrasound vibrations for 10 minutes.

The stainless steel coupons are manually dip coated in the prepared solutions. The coating process is repeated several times with intermediate drying at 200°C for 5 minutes. The coating times are the same for all the coupons in order to obtain similar coating thicknesses. Isothermal oxidation tests are carried out on both bare and coated coupons in dry air at 800°C and 1000°C. In addition, the exhibited oxide scale morphology is characterized by EDX and SEM means.

RESULTS AND DISCUSSION

Figure 2 shows the nanoparticle agglomerates obtained after processing and after dispersion by ultrasound treatment. Notice that the cluster sizes exhibit a significant size reduction due to ultrasound dispersion.

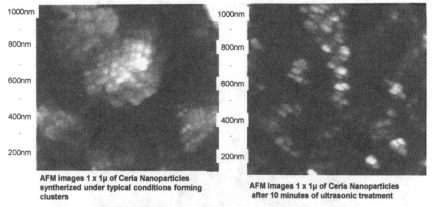

AFM Images 1 x 1µ of Ceria Nanoparticles syntherized under typical conditions forming clusters

AFM Images 1 x 1µ of Ceria Nanoparticles after 10 minutes of ultrasonic treatment

Figure 2. Atomic force microscope images of nanoceria particles alter (a) processing and (b) processing followed by ultrasound dispersion for 10 minutes.

Fig. 3. shows nanoceria coated (C) and bare (B) 316L stainless steel coupons exposed to various temperatures (800-1000°C) in dry air for 30 hours. From this figure it is evident that without the nanoceria coating the 316L stainless steel develops a dark and relatively thick scale. In contrast, the coated coupons do not seem to exhibit appreciable oxidation.

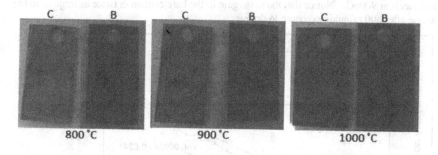

800 °C   900 °C   1000 °C

Figure 3. Optical images of coated (C) and bare 316L stainless steel coupons exposed to high temperature dry air for 30 hours.

In order to corroborate the beneficial effect of nanoceria coatings in 316L stainless steel, coated and bare coupons are exposed to 1000°C for 300 hrs in dry air. Fig. 4 shows the resultant scales after 300 hours exposure. Notice that the scale in the bare 316L stainless steel is totally spalling from the steel surface and that numerous blisters are present on the scale indicating that the scale is non-protective. In contrast, the scale developed in the nanoceria coated coupon is adherent and protective.

Figure 4. Optical images of coated (C) and bare 316L stainless steel coupons exposed to 1000°C dry air for 300 hours.

Figure 5 shows the thermal analysis results on weight gain versus time in bare and coated stainless steels at 900oC.   Notice that the mass gain in the bare coupon is twice as large as in the coated one after 600 minutes exposure to dry air.

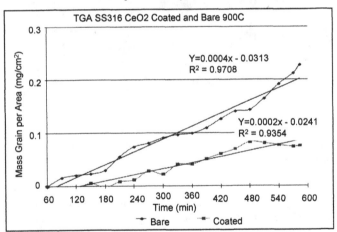

Figure 5. Thermogravimetric analysis of Coated and Bare coupons at 900°C for 6 hours in dry air. Mass gain per area vs. time.

From this figure it is evident that nanocrystalline cerium oxide coatings improve the high-temperature-oxidation resistance of the 316L stainless steel.

Although, the scales developed in nanoceria coated coupons provided significant protection, there is a concern related to how uniform the coatings are, and whether surface areas remain uncoated and thus exposed to appreciable oxidation. Fig. 6 shows the development of oxide islands on a coated 316L coupon. In addition, the oxide scale can be observed by optical means. EDX analyses indicate that the developed scale contains Ce and Cr, but it also includes Fe.

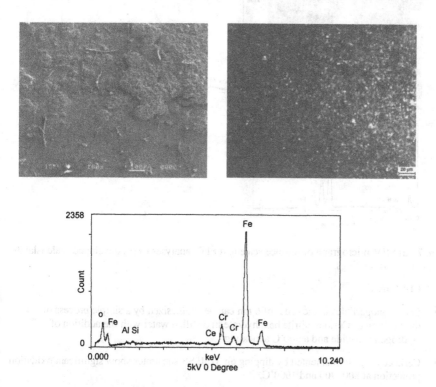

Figure 6. (a) SEM micrograph and (b) optical image of the developed surface scale, (c) EDX analysis of the surface scale.

A closer view to the developed scales is carried out by SEM means. Figure 7 shows oxide islands developed in the nanoceria coated coupon. EDX analyses indicated that these islands are Fe rich. In contrast, in the scale "valleys" the dominant elements are Cr and Ce. Apparently, in regions where effective nanoceria coatings are present, the scale is thin and uniform. However, in areas that are not well protected, a Fe-based scale develops in the form of islands.

The role of nanoceria coatings on the exhibited scale is investigated in a previous work [5]. From this work, it is found that nanoceria particles influence the resultant scale by promoting oxygen inward diffusion as the scale growth limiting step. In addition, they provide nucleation sites for the development of a fine grained scale and the resultant scale is a Cr oxide doped with Ce [5].

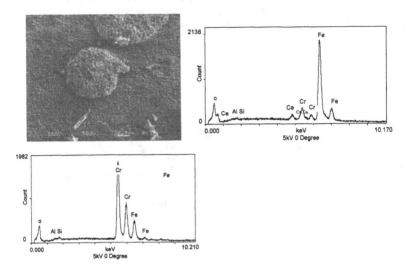

Figure 7. (a) SEM micrograph of Surface scale and EDX analyses of (b) developed scale islands and (c) their surroundings.

CONCLUSIONS

- Ceria nanoparticles in the order of 6 nm can be synthesized by a simple process of dissociation of Cerium Nitrite hexahydrate in distilled water with the addition of Hydrogen peroxide and a 300°C annealing.

- Ceria coatings implemented by dipping on 316L SS substrates show significant oxidation protection at 800, 900 and 1000°C.

- The developed scales are strongly influenced by the uniformity of the surface coatings.

REFERENCES

1. F. Czerwinski and W. Smeltzer, *Oxid. Met.*, **40** (1993), 503.
2. J. Shen, L. Zhou, and T. Li, *J. Mater. Sci.*, **33** (1998), 5815.
3. S. Patil, S. C. Kuiry, S. Seal and R. Vanfleet, *J. Nanopart. Res.*, **4** (2002), 433.
4. Zhonghua Wu et al, *J. Phys.: Condensed Matter*, **13** (2001), 5269.
5. Haying Zhang, Ph.D. Dissertation, Materials Department, University of Wisconsin-Milwaukee, 2006.

Mater. Res. Soc. Symp. Proc. Vol. 1276 © 2010 Materials Research Society

Preparation and Mechanical Characterization of a Polymer-Matrix Composite Reinforced with PET

J. Elena Salazar–Nieto[1], Alejandro Altamirano–Torres[1], Francisco Sandoval–Pérez[1] and Enrique Rocha–Rangel[2]

[1]Departamento de Materiales, Universidad Autónoma Metropolitana, Av. San Pablo No. 180, Col. Reynosa-Tamaulipas, México, D. F., 02200.

[2]Universidad Politécnica de Victoria, Avenida Nuevas Tecnologías 5902, Parque Científico y Tecnológico de Tamaulipas, Ciudad Victoria, Tamaulipas, 87137, México

ABSTRACT

In this study, polymer-matrix composites are fabricated by mixing liquid epoxy resin with 0, 15, 20 and 25 wt % of PET. PET is used as a reinforcement material since it can be recycled and this implies a beneficial environmental impact. After mixing, specimens are dried at room temperature during 24 h and then cured at 150°C during 0.5, 0.75 and 1 h. Then mechanical tests are performed. Experimental results obtained from the flexion test for 100 % epoxy resin and 15 % PET samples, without curing treatment show values of 30 and 21 MPa, respectively. Flexure strength values for the same samples but after curing treatment are: 56, 90, 32 MPa and 69, 64, 70 MPa, for 0.5, 0.75 and 1 h of treatment, respectively. These data show an important increase in the flexure strength for the sample reinforced with 15 % PET and curing time of 1h. This is most likely due to the behavior of PET's powders at this temperature and time. They can partially melt improving the adhesion to the polymeric matrix. For a curing time of 0.75h, this property decreases, due to the high porosity developed in the composite and the poor adhesion between polymeric matrix and reinforced material.

INTRODUCTION

Numerous studies exist on the properties of composite materials based on particles and synthetic fibers with thermo rigid resins, materials that are widely spread in the construction, transport and naval industries [1]. Recent investigations indicate that in the field of composite materials with polymeric matrix, research is orientated towards the substitution of the synthetic fibers as reinforcement materials. This is due to the disadvantages associated with the utilization of synthetic fibers (glass fiber, graze, etc.) i.e., high costs and the difficulty for recycling [2-5]. On the other hand, the worldwide objective to protect the environment has been reflected in an effort for avoiding pollution and the excessive exploitation of natural resources, which demands among other aspects the development of new materials, with better properties and harmless to the environment and in addition of low cost [6]. As response to this need, in the present work appears the creation of a material that covers these aspects, his environmental contribution will be take advantage of materials as degree bottle PET, which is considered to be a waste. With the results obtained from the mechanical characterization of this material, and the later comparison with existing similar materials, there will be decided the applications that this one could have.

EXPERIMENTAL PROCEDURE

PET's powder is prepared by mechanically grinding bottles, the final average particle size is 11 μm. This powder (0, 15, 20 and 25 wt %) is mechanically mixed with epoxy resin (MPT-XV, Polyform México). The mixture is poured into rubber molds with the form and dimensions of specimens for mechanical testing. Flexure and impact strength are tested by following ASTM standards. Samples are left to dry in the mold at room temperature during 24 h and then cured at 150 °C, for different times (0.5, 0.75 and 1h). Cured samples are characterized by measuring density and open porosity (Arquimedes method). The flexural strength is evaluated in a universal testing machine using a deformation speed of 0.127 cm/min and samples with the following dimensions: 0.4 cm thickness x 8 cm length x 2.4 cm width. The impact strength is evaluated on Izod type v-notched samples with the following dimensions 1.06 cm thickness, 7.3 cm length, 1 cm width and 0.2 cm notch depth. Impact test are carried out at different temperatures (-2 °C, 24 °C, 100 °C and 150°C). Surfaces of fractured samples are observed with the help of a stereomicroscopy. For each studied condition five different samples are evaluated.

RESULTS AND DISCUSSION

Density

Table I shows the results of density and porosity measurements evaluated by the Archimedes' method. From the obtained results, it can be seen that there is a clear increase of open porosity for relatively long curing times. This indicates that the samples do not completely solidify after pouring in molds and left for 24 h at room temperature. Most likely, air is dissolved during curing creating bubbles. However the sample containing 25 % PET shows a different behavior, as curing time increases, the porosity decreases. Although the initial values of porosity in this sample are high (even slightly higher than those in the other samples) after being cured, the effect is important. This suggests that in materials with a high content of PET (25 %) most likely the adhesion between resin and reinforcement material is bad, giving rise to the results already seen in this table. The final values of density in all cases are very similar.

Flexural strength

Figure 1 shows the results of three point flexure tests as a function of curing time for all fabricated samples. The original specimen (100% resin) increases its strength by increasing curing time, reaching a maximum at 0.75h. However at longer times, there is an important decrease in this property. On the other hand, samples containing 15, 20 and 25 wt % of PET show a relative increment of the flexure strength with curing time. The sample with 15 wt % PET presents the maximum flexure strength, whereas the sample containing 25 wt% PET renders the lower values of strength. As mentioned above this behavior probably has its origin in the degree of adhesion between epoxy resin and the reinforcement material.

Impact strength

Figure 2 shows impact testing results for samples cured for 0.5 h as a function of test temperature

Table I. Results obtained from measurements of density and open porosity in manufactured materials.

Wt % PET	Sample	Density (g/cm³)	Open porosity (%)
0	original	1.15	1.54
0	0.5 h	1.15	3.28
0	0.75 h	1.17	3.34
0	1.0 h	1.10	3.59
15	original	1.09	1.61
15	0.5 h	1.13	1.64
15	0.75 h	1.14	2.70
15	1.0 h	1.11	2.83
20	original	1.11	1.51
20	0.5 h	1.11	1.54
20	0.75 h	1.12	2.94
20	1.0 h	1.12	3.93
25	original	1.15	3.94
25	0.5 h	1.12	3.81
25	0.75 h	1.11	2.97
25	1.0 h	1.13	1.27

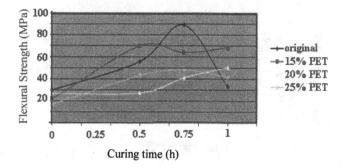

Figure 1. Three point flexure strength for all samples produced here as a function of curing time.

This figure shows that at 100°C for original and 25% PET samples there are increments in the impact strength, while in sample with 15% PET there is a slight decrement in impact strength when the material is tested between -5 and 25°C, at 150°C impact strength reaches a maximum of 18 N/m. The sample that shows better performance under these study conditions is one that contains 20 wt % PET, because its impact strength decreases from 16 to 10 N/m when sample is tested at -5 and 25°C respectively, however at 150°C this strength significantly increases reaching 35 N/m, which is the highest strength achieved by any of the samples analyzed here.

Figure 3 shows the behavior presented by the specimens subjected to the impact test for curing times of 0.75h, as a function of test temperature. Here in all cases the maximum impact strength is achieved when sample is tested at 150°C, the sample that absorbs more energy is one that contains 25% PET, although slightly below there is one containing 20% PET. At lower test

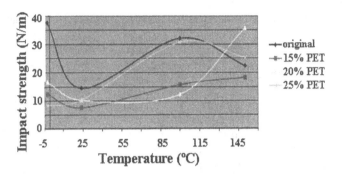

Figure 2. Impact strength according to the test temperature assessed in samples with different weight percentages of PET and a curing time of 0.5h at 150°C.

temperatures are behaviors somewhat erratic in all samples, however it can be commented that at 25°C the resistance displayed by the samples is greater than the resistance that is obtained at the same temperature in those samples cured for 0.5h (Figure 2).

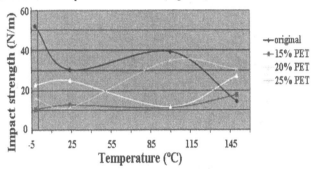

Figure 3. Impact strength according to the test temperature assessed in samples with different weight percentages of PET and a curing time of 0.75h at 150°C.

Figure 4 shows the behavior presented by the specimens subjected to the impact test for curing times of 1h, as a function of test temperature. This figure shows that the behavior of the original samples, 15 and 20% PET is very similar as it decreases slightly between -5 and 25°C and then increases to values of 33 and 37 N/m at 100°C to decrease below 20N/m when the test is performed at 150°C. Otherwise it is the one with the sample with 20% PET in which from 20 N/m at -5 ° C decreases to 15 N/m at 25 and 100 ° C, to increase sharply to 27 N/m at 150°C.

With regard to the study of the fracture surface of different studied samples under the above conditions, Figure 5 refers that when PET percentage increasing the type of fracture is changing, this means, surface is observed more opaque, indicating that the material is changing the fracture

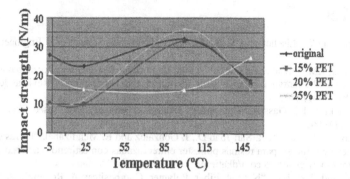

Figure 4. Impact strength according to the test temperature assessed in samples with different weight percentages of PET and a cured time of 1h at 150°C.

type from brittle to ductile. However, it is noteworthy that also observed an increase in closed pores, its mean internal pores of the material, that cause a drastic decrease in mechanical properties.

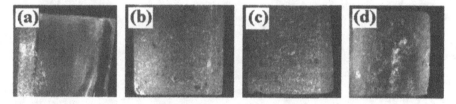

Figure 5. Fracture surfaces of studied samples; a) Original, b) 15 % de PET, c) 20 % de PET and d) 25 % de PET.

CONCLUSIONS

Curing times have a strong influence on the mechanical properties of the resulting materials. Being the combination of 0.75h at 150°C, the best curing conditions, because is where adhesion between resin and reinforcement material (PET) is better achieved, with the exception of the sample with 15% PET. The mechanical properties such as toughness and impact strength of composite material fabricated, are higher than the material made only with epoxy resin, indicating a reinforcing effect of PET in the epoxy resin. From the values of flexure strength, toughness and impact strength, it follows that a potential application of the compound, is to replace automotive parts such as fenders, which currently are made of resin and fiberglass, as well as being able to produce signs: road, safety and hygiene, ecology and civil protection, among others.

REFERENCES

1. S.E. Buck, D.W. Lischer and S. Nemat-Nasser, Material Science and Engineering, **A317** (2001) 128.

2. J.M. Arribas, J.M. Navarro and C. Rial, "Compuestos de polipropileno reforzado con fibras Vegetales; Una alternativa ecológica para la industria del automóvil". Revista de Plásticos Modernos, **81** (2001) 467.

3. A.K. Bledzki and J. Gassan. "Composites reinforced with cellulose based fibers". Elsevier Science Ltd., (1999).

4. C.F. Jasso, H. Hernández, R. San-Juan D., J. González and E. Mendizábal, "Fibras celulósicas como agentes de refuerzo para resinas poliéster entrecruzadas con estireno o acrilato de butilo". CUCEI. http://www.geocities.com/dkatime.

5. D. Nabi and J.P. Jog, "Natural Fiber Polymer Composites: A Review", Adv. Polym. Technology, **18** (1999) 351.

6. J. Madera Santana, M. Aguilar Vega and F. Vázquez Moreno F., "Potencial de las fibras naturales para su uso industrial", Ciencia Ergo Sum, Noviembre 2000, Toluca, México.

Mater. Res. Soc. Symp. Proc. Vol. 1276 © 2010 Materials Research Society

Micromechanical Models of Structural Behavior of Concrete

Ilya Avdeev[1], Konstantin Sobolev[2], Adil Amirjanov[3], Andrew Hastert[1]

[1]Department of Mechanical Engineering, University of Wisconsin-Milwaukee, WI 53211, U.S.A.
[2]Department of Civil Engineering, University of Wisconsin-Milwaukee, WI 53211, U.S.A.
[3]Department of Computer Engineering, Near East University, Nicosia, TRNC, Mersin 10, TURKEY

ABSTRACT

A three-dimensional numerical model capable of predicting structural behavior of concrete under various loading conditions is developed. Concrete, as a composite material, is represented by the mechanically strong aggregates of various shapes and sizes incorporated into a cement matrix. The most important aspect of concrete modeling involves an accurate representation of the spatial distribution of the aggregate particles.

A micromechanical heterogeneous model based on prescribed spatial distribution of aggregates is developed. This model allows to compute the effective material properties of concrete using a representative cell homogenization approach. The results of numerical analysis of this model are compared to the models of particulate composite material.

INTRODUCTION

The macromechanical properties and behavior of particulate composite materials (strength, modulus of elasticity, creep, and shrinkage), such as portland cement and asphalt concrete mixtures, greatly depend on the properties of their main constituent: the aggregates [1]. In addition to material properties, the most important parameters that affect the macro-mechanical behavior include the packing density, compaction degree, and particle size distribution of aggregates. Better packing distributions improve the behavior of concrete [1, 2]. While there are currently several simple models that can predict the macro-mechanical properties of aggregate based composites, there is a need to create one that takes into account the size distribution and spacing of aggregate particles (Figure 1) [4, 6].

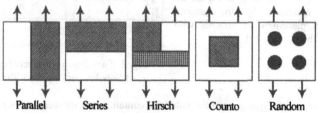

Parallel   Series   Hirsch   Counto   Random

Fig. 1. Various simplified matrix/aggregate models for macro-mechanical behavior, [4,6].

The elastic modulus of the *composite*, E_c is based on these models summarized in Table I.

Table I. Simplified aggregate particle models

Parallel [4]	$E_c = E_p V_p + E_m V_m$	Series [4]	$\frac{1}{E_c} = \frac{V_p}{E_p} + \frac{V_m}{E_m}$
Hirsch [4]	$\frac{1}{E_c} = X\left(\frac{1}{V_p E_p + V_m E_m}\right) + (1-X)\left(\frac{V_p}{E_p} + \frac{V_m}{E_m}\right)$	Counto [4]	$\frac{1}{E_c} = \frac{1 - V_p^{\frac{1}{2}}}{E_m} + \frac{1}{\left(\frac{1 - V_p^{\frac{1}{2}}}{V_p^{\frac{1}{2}}}\right)E_m + E_p}$

Random (Maxwell Model) [6]	$E_c = E_m \dfrac{\left[1 + 2V_p\left(\frac{E_p}{E_m} - 1\right)/\left(\frac{E_p}{E_m} + 2\right)\right]}{\left[1 - V_p\left(\frac{E_p}{E_m} - 1\right)/\left(\frac{E_p}{E_m} + 2\right)\right]}$

In these equations V_p is the volumetric proportion of *particulate*, E_p is the elastic modulus of the *particulate*, V_m is the volumetric proportion of the *matrix*, and E_m is the elastic modulus of the *matrix*.

Sobolev Random Particle Packing Algorithm

Given the relationship between input parameters and output material performance, it is of great importance that the input parameters, including spacial distribution of aggregates, be highly tunable. The Sobolev packing algorithm is used to create a parametric aggregate dispersion for use in mechanical simulation (Figure 2) [1-3]. The packing size distribution is highly tunable and parametric-based which allows for future coupling of parameters and output values for modeling of real-scale particulate composites. The packing algorithm begins with the random generation of a center for the first large sphere ($d_1 = D_{max}$ in Figure 2). The method continues by the random generation and placement of spheres with diameter d_i, such that $D_{max} \geq d_i \geq D_{min}$.

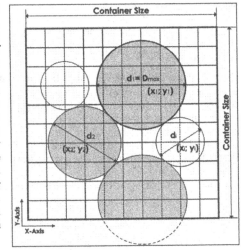

Fig. 2. Two-dimensional representation of Sobolev packing algorithm [1].

The center of the sphere must lie within the container and spheres may not overlap. Any spheres that do not fit these criteria are discarded.

After achieving a certain number of packing trials D_{min} is reduced using the coefficient K:

$$D_{min} = \frac{D_{max}}{\left(1 + 10^K\right)^N}$$

where D_{min} is the minimum diameter, D_{max} is the maximum diameter, K is the reduction coefficient, and N is the number of trials. This method and algorithm yields a simple and pseudo-dynamic packing system that produces randomly packed aggregates in a given container. This method is further expanded to allow for coefficient-driven particle spacing, allowing for a more realistic suspension model to be generated:

$$ r = \frac{d}{(1 + \frac{k_{del}}{(1+10^S)^m})} $$

where r is the radius of the new sphere, d is a minimum distance to the surface of any already packed spheres, k_{del} is an initial coefficient to provide the separation between spheres, s is a coefficient to change the size of separation, and m is the number of steps to change the size. This process yields a favorable result for the modeling of aggregate based composites when the spacing algorithm is applied.

Finite Element Micromechanical Model

When considering the length scale and controlled volumetric input, the most appropriate means of mechanical analysis is the Finite Element Method (FEM). The FE model used in numerical experiments is based on the 3-D solid geometry (cement matrix and a set of aggregates) generated with the random packing algorithm. Isoparametric second-order 3-D solid elements are used in this study [7]. Each element has 20 nodes with three degrees of freedom per node (nodal translational displacements). The models used in numerical experiments consisted of 279,896 elements and 378,089 nodes. Linear elastic material models are used for both cement and aggregate phases. Also, structural deformations are assumed to be relatively small so that the non-linear effects could be ignored.

NUMERICAL EXPERIMENT

The first challenge is found in creating a simplified spherical particulate composite three-dimensional model. This model must incorporate a random distribution of aggregate particles, but the particle size distribution and packing density must remain parametric. For our purposes, the Sobolev packing algorithm provides good results for the FEM input model (Figure 3). Once the input parameters are established, five similar random aggregate packing models are created using the same parameters as described in Ref. [1] (D_{max} = 5000, N = 10^6, K = 1, k_{del} = 1.5, S = 1.001). This offers a five-time replicated experiment for verification of computational results. Based on these computational experiments, the average packing density of particles is 25% (Table II). These volumes of aggregates represent only the coarsest aggregate fraction corresponding to the particle size of 10-25 mm. The 300 volumetric aggregate particles are suspended in a matrix and a three-dimensional solid mesh is generated. Material properties for the matrix and aggregate are applied. Young's Modulus of concrete varies in the range 20-50 GPa and that of common aggregates in the range 30-70 GPa. Thus, four numerical experiments are conducted using the the extreme values of Young modulii for both concrete and aggregate. The Poisson's ratio used in the calculations is 0.2 for both aggregate and matrix material. This ensures a control for one of the main variables of structural macromechanical properties and behavior.

137

The following boundary conditions are applied: one XY planar face is constrained in the Z-direction and all nodes are coupled in the X and Y directions, opposite XY planar face is loaded with a constant pressure that limits experiment to the linear/elastic range of analysis, and all nodes on this face are coupled in the X and Y directions. The structure is loaded like a simple uniaxial compression test with no friction on the planar surfaces applying pressure. This allows for a simulation procedure that mimics the laboratory experiment on concrete composites used for further verification of simulated models.

In total, twenty simulations are performed: four material conditions across five spacial packing distributions obtained at the same packing parameters. Von Mises stress, strain, and macromechanical modulus of elasticity is computed for the composite structure. With a uniform applied pressure, the most interesting output of simulated experiments is the structural strain [5].

$$\lambda = stress/strain$$

For verification of method and experimental inputs, the multi-axial strain and calculated Poisson's ratio is observed to closely resemble the input control ratio (Table II). With post-simulation macromechanical Poisson's ratio controlled, the results can clearly focus on observed elastic modulus (Table II). Also, the modulus of elasticity is calculated using the various simplified methods discussed before. The results are compared to the calculated values found by numerical (FEM) analysis (Table III).

Table II. Simulated compression test results

			Experiment 1		Experiment 2		Experiment 3		Experiment 4	
			Elastic Modulus (Pa)	Poisson Ratio	Elastic Modulus (Pa)	Poisson Ratio	Elastic Modulus (Pa)	Poisson Ratio	Elastic Modulus (Pa)	Poisson Ratio
Common Input Properties	Aggregate		3.00E+10	0.2	3.00E+10	0.2	7.00E+10	0.2	7.00E+10	0.2
	Matrix		2.00E+10	0.2	5.00E+10	0.2	2.00E+10	0.2	5.00E+10	0.2
		Vol. %								
Solved Properties	Pack 1	25.14	2.08E+10	0.2004	4.76E+10	0.1992	2.25E+10	0.2005	5.17E+10	0.2004
	Pack 2	24.89	2.08E+10	0.2004	4.77E+10	0.1992	2.23E+10	0.2006	5.16E+10	0.2004
	Pack 3	24.88	2.08E+10	0.2003	4.76E+10	0.1993	2.25E+10	0.2002	5.17E+10	0.2003
	Pack 4	24.89	2.08E+10	0.2003	4.76E+10	0.1993	2.25E+10	0.2003	5.17E+10	0.2003
	Pack 5	24.58	2.08E+10	0.2004	4.76E+10	0.1993	2.24E+10	0.2005	5.17E+10	0.2003
	Avg. (Pa)		2.08E+10	0.2004	4.76E+10	0.1993	2.25E+10	0.2004	5.17E+10	0.2003
	Stand. Dev.		2.16E+07	4.21E-05	5.82E+07	4.10E-05	6.75E+07	1.70E-04	4.42E+07	3.56E-05
	Variance		4.69E+14	1.77E-09	3.39E+15	1.68E-09	4.56E+15	2.90E-08	1.96E+15	1.27E-09

SIMULATION RESULTS AND DISCUSSION

Similarity of Composite and Matrix Moduli

The final composite elastic modulus is measured from post-process simulated solutions. The resulting composite elastic modulus for each experiment closely resembled that of the original matrix elastic modulus value. While none of the simplified aggregate composite models provide these exact results, they are within the logical range given the low aggregate volumetric ratio (Table II). This further substantiates a need for a new parametric model that accounts for aggregates of mid and fine particle sizes.

Table III. Comparison of various simplification models to numerical results

Comparing Elastic Modulus Results (GPa)							
	Experiments						
	1		2		3		4	
	Agg.	Mat.	Agg.	Mat.	Agg.	Mat.	Agg.	Mat.
	30	20	30	50	70	20	70	50
Approach								
Parallel	20.95		48.10		24.75		51.90	
Series	20.65		47.02		21.46		51.40	
Hirsch	20.80		47.55		22.99		51.65	
Counto	20.86		47.92		23.10		51.75	
Random	20.83		47.84		22.71		51.70	
Numerical	20.82		47.61		22.45		51.71	

Further simulations on composites with a large number of particulate inclusions will be required. The experimental input range will be expanded to include a full range of particle size distributions as well as particle spacing to imitate the range found in "real" concrete.

Interstitial Space Stress Concentration

The macromechanical model can be optimized for mechanical properties, like elastic modulus, for a given range of aggregate and matrix properties. To further the capabilities of the model for industrial use, non-linear analysis will be performed on a finer meshed particulate composite model. The most interesting aspect of this line of experimentation will focus on the initiation and arrest of cracks in the interstitial space around aggregates (Figure 4). The model will be expanded for parametric relationship between particle size distribution, aggregate density, particle spacing and crack propagation.

CONCLUSIONS

Implementation of the random packing algorithm proved to be consistent between the five iterations. Coupling of the random packing algorithm with the finite element method of analysis produced results similar to that of the simplified models discussed in Figure 1, but it can be assumed that with variation in packing density, aggregate size distribution, and separation coefficients the results will deviate. Resulting macromechanical elastic modulus as measured through numerical analysis resembles the Random (Maxwell) results most closely. The resulting measured elastic modulus in each numerical experiment closely matched that of the primary constituent – the matrix material. While the coupled method has proved to produce favorable results, further verification of the method with concrete specimen experimentation is required. In the future, this experimentation will be augmented with a wide range of separation coefficients, input material properties, and packing densities.

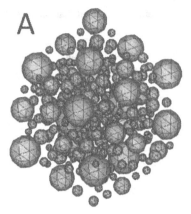

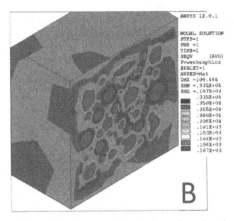

Fig. 3 (A). Meshed suspension aggregates

Fig. 4 (B). Cross-sectional planar face showing stress concentrations around aggregate/matrix interstitial space

ACKNOWLEDGMENTS

Authors would like to acknowledge the financial support from the Research Growth Initiative (RGI), National Science Foundation (NSF), Stipends for Undergraduate Research Fellows (SURF), ANSYS Institute of Industrial Innovation at UW-Milwaukee (AIII), and CONACYT. Special appreciation is conveyed to Mir Zunaid Shams for his contribution in finite element modeling.

REFERENCES

1. K. Sobolev and A. Amirjanov, Application of genetic algorithm for modeling of dense packing of concrete aggregates, Const. and Build. Mater. 24, 8, 1449-1455 (2010).
2. K. Sobolev and A. Amirjanov, A simulation model of the dense packing of particulate materials, Adv. Powder Technol. 15, 365-376 (2004).
3. A. Amirjanov and K. Sobolev, Fractal properties of Apollonian packing of spherical particles, Model. Simul. Mater. Sci. Eng. 14, 789-798 (2006).
4. J. Young, S. Mindess, R. Gray, A. Benhur, *The Science and Technology of Civil Engineering Materials*, 179-188 (1998). U.S.: Prentice Hall.
5. C. Hartsuijker, J.W. Welleman, *Engineering Mechanics Volume 2*, (2001). U.S.: Springer.
6. Y. Ohama, Principle of Latex modification and some typical properties of Latex-modified mortar and concrete, ACI Mater. Jour. 84, 6, 511–518 (1987).
7. ANSYS 12.0 documentation 2009 ANSYS, Inc., Canonsburg, PA, USA

Mater. Res. Soc. Symp. Proc. Vol. 1276 © 2010 Materials Research Society

Performance of Cement Systems with Nano- SiO$_2$ Particles Produced Using Sol-gel Method

Konstantin Sobolev[1], Ismael Flores[1], Leticia M. Torres[2], Enrique L. Cuellar[3], Pedro L. Valdez[2], and Elvira Zarazua[2]

[1] Department of Civil Engineering, CEAS, University of Wisconsin-Milwaukee, 3200 N. Cramer St., Milwaukee, WI 53211, USA
[2] Facultad de Ingeniería Civil, Universidad Autónoma de Nuevo León S/N, San Nicolás de los Garza, N.L. 66400, México
[3] Facultad de Ingeniería Mecánica y Eléctrica, Universidad Autónoma de Nuevo León S/N, San Nicolás de los Garza, N.L. 66400, México

ABSTRACT

The reported research examines the effect of 5-70 nm SiO$_2$ nanoparticles on the mechanical properties of nanocement materials. The strength development of portland cement with nano-SiO$_2$ and superplasticizing admixture is investigated. Experimental results demonstrate an increase in the compressive strength of mortars with SiO$_2$ nanoparticles. The distribution of nano-SiO$_2$ particles within the cement paste plays an essential role and governs the overall performance of these products. Therefore, the addition of a superplasticizer is proposed to facilitate the distribution of nano-SiO$_2$ particles. The application of effective superplasticizer and high-speed dispergation are found to be very effective dispersion techniques that improve the strength of superplasticized portland cement mortars, reaching up to 63.9 MPa and 95.9 MPa after aging during 1 and 28 days, respectively. These values compare favorably with the observed compressive strengths of reference portland cement mortars of 53.3 MPa and 86.1 MPa. It is concluded that the effective dispersion of nanoparticles is essential to obtain the composite materials with improved performance.

INTRODUCTION

Recent research in cement and concrete has focused on the investigation of the structure of cement-based materials and their fracture mechanisms [1-5]. The application of Atomic Force Microscopy (AFM) for the investigation of the "amorphous" C-S-H gel revealed that at the nanoscale this product has an ordered structure [6]. Better understanding the nano-structure of cement based materials helps to control the processes related to hydration, strength development, fracture, and corrosion. For instance, the development of materials with new properties such as self-cleaning, discoloration resistance, anti-graffiti protection, and high scratch and wear resistance, is important for many construction applications [7-11]. Another example is related to the development of new superplasticizers for concrete, such as Sky, which is based on polycarboxylic ether (PCE) polymer. This product is developed by BASF with a nano-design approach targeting the extended slump retention of concrete mixtures [12].

Major improvements in concrete performance have been achieved by addition of superfine particles, for example: fly ash, silica fume, metakaolin and now nanosilica. The optimal performance of these systems is attributed to the high-density continuous packings of the binder constituents that are realized at high fluidity levels with the help of effective superplasticizers [3-5]. For example, silicon dioxide nanoparticles (nanosilica, nano-SiO$_2$) proved to be a very effective additive to polymers for improving strength, flexibility, and durability. Nano-SiO$_2$ can be used as an additive to improve the workability and the strength of high-performance and self-compacting concrete [3-5, 13-15].

The compressive and flexural strength of cement mortars increase with the addition of Fe$_2$O$_3$ and SiO$_2$ nanoparticles [15, 16]. It is found that increasing the nano-SiO$_2$ dosage improves the strength of mortars to values higher than those observed in mortars with additions of silica fume. Collepardi et al. investigated self-compacting concretes with low-heat development [13-14]. Mineral additives such as ground limestone, fly ash and ground fly ash are used to control the heat of hydration. Nanosilica (particle size between 5 and 50 nm) has been used as a viscosity modifying agent at a dosage of 1-2% of cementitious materials. The best performance is shown by concrete with ground fly ash, 2% nanosilica and 1.5% of superplasticizer. This concrete has the highest compressive strength of 55 MPa after aging during 28 days and an excellent behavior in a fresh state: low bleeding, perfect cohesiveness, better slump flow and very little slump loss. Furthermore, nano-binders composed of nanosized cementitious material, pozzolanic nanoparticles and a finely ground mineral additive-portland cement mixture have been proposed [3, 4, 5].

Sol-gel method to produce nanoparticles

Relatively small quantities (less than 1 wt%) of nanosized materials are sufficient to improve the performance of nanocomposites [3-5, 17]. Yet, the commercial success of nanomaterials depends on the ability to manufacture these materials in large quantities and at a reasonable cost relative to the overall effect of the nanoproduct. Nanomaterial production technologies with potential industrial applicationn include plasma arcing, flame pyrolysis, chemical vapour deposition, electrodeposition, sol-gel synthesis, mechanical attrition and the use of natural nanosystems [18]. Among chemical technologies, sol-gel synthesis is one of the widely used "bottom-up" production methods for nano-sized materials, such as nano-silica (Figure 1). The process involves the formation of a colloidal suspension (sol) and gelation of the sol to form a network in a continuous liquid phase (gel). Usually, trymethylethoxysilane or tetraethoxysilane (TMOS/TEOS) is applied as a precursor for synthesizing nanosilica. The sol-gel formation process can be simplified to few stages [3, 18]:

1. Hydrolysis of the precursor;
2. Condensation and polymerization of monomers to form the particles;
3. Growth of particles;
4. Agglomeration of particles, followed by the formation of networks and, subsequently, gel structure;
5. Drying (optional) to remove the solvents and thermal treatment (optional) to remove the surface functional groups and obtain the desired crystal structure.

There are a number of parameters that affect the process, including pH, temperature, concentration of reagents, H$_2$O/Si molar ratio (between 7 and 25) and type of catalyst [18]. When precisely

142

executed, this process is capable of producing perfectly spherical nanoparticles of SiO_2 within the size range of 1–100 nm. The chemical reaction of nanosilica synthesis can be summarized as:

$$n\,Si(OC_2H_5)_4 + 2n\,H_2O \xrightarrow[NH_3]{C_2H_5OH} n\,SiO_2 + 4n\,C_2H_5OH$$

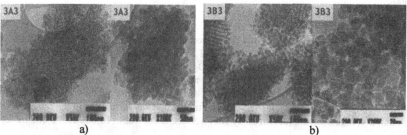

Fig. 1. The morphology and particle size distribution of nano-SiO_2 (TEM) synthesized by the sol-gel method using a molar ratio of TEOS/Etanol/H_2O of 1/6/24 at pH=2 (a) and pH=9 (b).

EXPERIMENTAL

The SiO_2 nanoparticles used in this work are synthesized using the sol-gel method using tetraethoxysilane (98% TEOS, supplied by Aldrich) as a precursor. The reaction is carried out in alkaline or acid media using ammonia as a catalyst (ammonia solution at pH = 9). This methodology produces nano-SiO_2 with particle size in the range of 5–100 nm depending on the experimental conditions: ethanol-to-TEOS molar ratios 6 or 24 and water-to-TEOS ratios of 6 or 24, as presented in Table I. Portland cement (NPC, conforming to ASTM Type I, supplied by CEMEX) and silica fume (SF, supplied by Norchem) are used in the experimental program. In addition, commercial nano-SiO_2 admixture Cembinder-8 (CB8, 50% water suspension , supplied by Eka Chemicals) is used as reference material. After evaluation of performance, commercial polyacrilate/polycarboxylate superplasticizer (PAE/SP, 31% concentration, supplied by Handy Chemicals) is used as a modifying admixture. Prior to application in mortars, SP is premixed with 18 wt% of tributyl-phosphate (99% TBP, supplied by Aldrich) in order to compensate for the air-entraining effect of PAE. Graded Ottawa sand (ASTM C778) is used as a fine aggregate in all tested mortars. Distilled water is used for the preparation of mortars.

The synthesized nano-SiO_2 particles are characterized by the X-ray diffraction (XRD, Bruker AXS D8), transmission electron microscopy (TEM, JEOL-2010), scanning electron microscopy (SEM, JEOL), and nitrogen absorption (BET; Quantachrome Nova E2000), as reported in Table I.

The performance of nano-SiO$_2$ based mortars (at a nano-SiO$_2$ dosage of 0.25 wt% of the binder; water-to-cement ratio (W/C) of 0.3 and sand-to-cement ratio (S/C) of 1) is compared with the properties of two reference mixtures: plain mortar (W/C of 0.3 and S/C of 1) and super-plasticized mortar (SP dosage of 0.1%). Relevant ASTM standards are used for evaluation of mortars flow (ASTM C1437) and compressive strength (ASTM C109, using 50x50x50 mm specimens).

RESULTS AND DISCUSSION

According to the results of X-ray diffraction, the obtained nano-SiO$_2$ is a highly amorphous material with predominant particle size of 1-2.5 nm [3]. Figures 1a and 1b illustrate the morphology and particle size distribution of nano-SiO$_2$ particles produced using a TEOS/ethanol/water molar ratio of 1/6/24 in acid and alkaline reaction media, respectively.(). The obtained nano-SiO$_2$ particles are represented by highly agglomerated xerogel clusters with sizes bwetween 0.5 and 10 μm. The particles within the clusters have sizes in the range 5-70 nm and obtained xerogels are characterized by BET surface areas from 116,000 to500,000 m^2/kg.

Table I. Design and properties of nano-SiO$_2$

Specimen Type*	Molar Ratio TEOS/Etanol/ H2O	Reaction Time, hours	Particle Size (TEM), nm	Surface Area (BET), m^2/kg
1B3	1/ 24 / 6	3	15-65	116,000
2B3	1/ 6 / 6	3	30, 60-70	145,000
3B3	1/ 6 / 24	3	15-20	133,000
4B3	1/ 24 / 24	3	5	163,000
1A3	1/ 24 / 6	3	5	510,100
2A3	1/ 6 / 6	3	<10	263,500
3A3	1/ 6 / 24	3	<10, 17	337,100
4A3	1/ 24 / 24	3	5	382,200

* Sample coding: ⌐ First number denotes molar ratio combination as per as experimental matrix
ABC — Last number corresponds to the reaction time in hours
 ∟ Letter – denotes reaction media: A- for acid and B- for base

The majority of obtained nano-SiO$_2$ particles (used at a dosage of 0.25%) reduce the flow of plain mortars to some extent [3]. In cases whre the superplasticizer (PAE) is applied at a dosage of 0.1%, the major part of the obtained nano-SiO$_2$ did not reduce the fluidity relative to that of the reference superplasticized mortar (Fig. 2). The performance of nano-SiO$_2$ depends on the conditions of synthesis (i.e., molar ratios of the reagents, type of the reaction media, pH and the duration of the reaction) [3]. The best nano-SiO$_2$ products with particle size ranging from 5 to 20 nm are synthesized at highest molar concentrations of water (specimens 3B and 4B). The addition of nano-SiO$_2$ to plain portland cement mortars at a dosage of 0.25% (by the weight of cementitious materials) improves the 1-day strength by up to 17%. Early strength (up

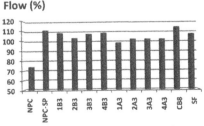

Flow (%)

Fig. 2. Mortars flow: reference vs. mortars with nano-SiO$_2$.

to three days) is also improved with the majority of obtained nano-SiO_2. The addition of nano-SiO_2 improves the 28-day compressive strength of plain mortars by 10% [3]. The distribution of nano-SiO_2 particles within the cement paste is an important factor governing the performance of these products [3]. Therefore, the dispersion of nanoparticles is essential to obtain a composite materials with improved properties. The application of PAE superplasticizer, ultrasonification and/or high-speed mixing is proven to distribute nano-SiO_2 [3].

According with the results of the experiment, the strength of superplasticized cement mortars with nano-SiO_2 is improved (Fig. 3). The addition of 0.1% of superplasticizer and 0.25% of nano-SiO_2 (specimen 4B3, manufactured using a TEOS/Ethanol/ H_2O molar ratio of 1:24:24 in alkaline medium) results in 1-day strength of 63.9 MPa. This value represents a 20% increase with respect to plain cement mortar (Fig. 3) and a 16% strength improvement with respect to superplasticized cement mortar. The strength of mortars with 4B3 at the age of 3 to 7 days is similar to the strength of superplasticized mortar. Only a slight improvement in 28-day strength, about 4% with respect to superplasticized mortar, is observed in this specimen reaching 95.9 MPa. The addition of 0.25% of nano-SiO_2 (specimen 1A3, manufactured using a TEOS/ Ethanol/H_2O ratio of 1:24:6 in acid medium) results in 1-day strength of superplasticized mortar of 61.7 MPa, about 16% increase with respect to plain cement mortar (Fig. 3); this value also represents a 12% strength improvement with respect to superplasticized cement mortar. The superplasticized cement mortar that exhibited the best performance and 1 day compressive strength is prepared with nano-SiO_2 particles manufactured using a TEOS/Ethanol/H_2O ratio of 1:6:24 in acid medium (specimen 3A3 in Fig. 1). Compressive strengths of 75.1, 80.3 and 96.5 MPa are observed at 3, 7 and 28 days, respectively. Application of SF and commercial nano-SiO_2 (CB8) to superplasticized mortars exhibit similar behavior. However, the increase in 1-day strength is only 8–9% with respect to superplasticized mortar. These additives slightly reduce the 28-day strength of superplasticized mortars. It can be expected that the application of wider range of particle sizes and better dispersion of nano-SiO_2 can further improve the performance of engineered nano-particles [3].

Fig. 3. Compressive strength of investigated mortars at different ages

CONCLUSIONS

The efficiency of nanoparticles such as nano-SiO_2 depends on their morphology and genesis, as well as on the application of superplasticizer and additional treatment options such as ultrasonification. With the sol-gel method, it is possible to manufacture a

Compressive Strength, MPa

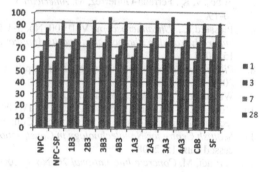

wide range of nanoparticles with engineered properties such as particle size, porosity and surface conditions. It is demonstrated that all synthesized nano-SiO_2 at a small dosage, 0.25% can improve the compressive strength of portland cement mortars. The major problem related to application of nano-SiO_2 is related to the agglomeration of nanoparticles and their even distribution within the cement matrix. The application of effective superplasticizer and high-speed mixing are found to be very effective disagglomeration techniques that improve the strength of superplasticized portland cement systems. It is observed that the best early and long-term strength is achieved by mortars based on nano-SiO_2 with particle size of less than 20 nm which are synthesized at higher water-to-TEOS molar ratio.

Further research is required to modify the sol-gel method to avoid the formation of dense xerogel agglomerates (such as the development of nano-SiO_2 products in liquid state, application of surfactants, ultrasonification and microwave drying), and to achieve a better dispersion of nano-SiO_2. Further investigation is necessary to quantify the effects of synthesized nano-SiO_2 on the hydration of portland cement-based systems.

ACKNOWLEDGEMENTS

The reported study is performed under the research grants of CONACYT 46371 and PROMEP 103.5/07/0319. The financial support of this project from CONACYT, PROMEP and PAICYT (Mexico) is gratefully acknowledged. Support of CEMEX, Handy Chemicals and Eka Chemicals for supply of cement materials, superplasticizers and nano-products is acknowledged.

REFERENCES

1. Gann, D. SPRU, University of Sussex, 2002.
2. Trtik, P., Bartos, P.J.M, *Proceeding of the 2nd Anna Maria Workshop: Cement & Concrete: Trends & Challenges*, 2001, pp. 109-120.
3. Sobolev, K.. Progress report, CONACYT, Mexico, 2006.
4. Sobolev, K., Ferrada-Gutiérrez, M.. *American Ceramic Society Bulletin*, No. 10, 2005, pp. 14-17.
5. Sobolev, K., Ferrada-Gutiérrez, M. *American Ceramic Society Bulletin*, No. 11, 2005, pp. 16-19.
6. Plassard, C., *Ultramicroscopy*, Vol. 100, No. 3-4, 2004, pp. 331-338.
7. *Super-Hydrophilic Photocatalyst and Its Applications* http://www.toto.co.jp Accessed June 15, 2008.
8. *Pilkington Activ™ - Self-cleaning Glass*, http://www.pilkington.com Accessed June 15, 2008.
9. Watanabe, T. US Patent 6294247, 2001.
10. Roco, M.C., Williams, R.S., Alivisatos, P.. IWGN Report on Nanotechnology Research Directions, National Science and Technology Council, Committee on Technology, 1999.
11. *Scottish Centre for Nanotechnology in Construction Materials*, www.nanocom.org Accessed June 15, 2008.
12. Corradi, M. *Concrete International* 26, No. 8, 2004, pp. 123-126.

13. Collepardi, M.. In: *Proceedings of the International Conference - Challenges in Concrete Construction - Innovations and Developments in Concrete Materials and Construction,* Dundee, UK , 2002, pp. 473 - 483.
14. Collepardi, *Proceedings of 8th CANMET/ ACI International Conference on Fly Ash, Silica Fume, Slag and Natural Pozzolans in Concrete,* SP-221, Las Vegas, USA, 2004, pp. 495-506.
15. Li, H. *Composites: Part B,* No. 35, 2004, pp. 185-189.
16. Li, G., *Cement and Concrete Research,* No. 34, 2004, pp. 1043-1049.
17. Bhushan, B. *Handbook of Nanotechnology,* Springer, 2004.
18. Wilson, M. Chapman & Hall/CRC, 2000.

Mater. Res. Soc. Symp. Proc. Vol. 1276 © 2010 Materials Research Society

Preparation and Mechanical Characterization of Composite Material of Polymer–Matrix with Starch Reinforced with Coconut Fibers

Yaret G. Torres-Hernández[1], Alejandro Altamirano-Torres[1], Francisco Sandoval-Pérez[1] and Enrique Rocha Rangel[2]

[1]Departamento de Materiales, Universidad Autónoma Metropolitana Av. San Pablo No. 180, Col. Reynosa-Tamaulipas, México, D. F., 02200.

[2]Universidad Politécnica de Victoria, Avenida Nuevas Tecnologías 5902, Parque Científico y Tecnológico de Tamaulipas, Ciudad Victoria, Tamaulipas, 87137, México

ABSTRACT

In this work the synthesis and mechanical characterization of a polymer matrix composite is reported. An epoxy resin is used as matrix with addition of starch and coconut fibers as reinforcement. Vickers hardness and impact tests are used for mechanical characterization. Starch is used to promote degradability of the polymer matrix with clear benefits for the environment. Natural fibers have been used for reinforcing the composite materials. Natural fibers have several advantages such as price, low density and relatively high mechanical properties, they are also biodegradable and non abrasive In this investigation, the composite material samples are fabricated with epoxy resin, 5, 10, 15 wt % of starch and 5, 10 wt % of coconut fibers with the help of silicon molds which have the dimensions and geometry according to ASTM Standards for make Impact and Vickers hardness tests. The obtained results show that increases in the amount of coconut fibers cause an enhancement of the mechanical properties of the material, due to a good adhesion between the polymeric matrix and the natural fibers.

INTRODUCTION

Due to the concern in the world related to environmental pollution, there has been a search for substitute materials with a reduced degradation time as compared to petroleum derived products without decreasing their mechanical properties. There are several studies about it [1, 2], for example using composites and natural fibers as reinforcements, normally finding an improvement in their properties and at the same time increasing their degradability. The use of a fiber is traditionally important because if added properly, provides greater mechanical strength to the final product [3]. On the other hand, there have been several projects on plastics reinforced with natural fibers in various countries, as for example the case of fiber reinforced plastics lignocellulosic [4] where a polypropylene matrix reinforced with hemp fibers is evaluated in terms of properties such as Young's modulus and impact resistance. It has been found that this material has a Young's modulus comparable to that of the same polymer reinforced with glass fiber. In addition there are several economic benefits. In this work the addition of starch has the objective to accelerate degradation of the composite material, an effect already observed in previous works [5, 6] and the use of coconut fiber as a mechanical reinforcement.

EXPERIMENTAL PROCEDURE

For the fabrication of the composite material epoxy resin (MPT-XV, Mexico) is used, with 5, 10, 15 wt % of starch, then coconut fibers (5 and 10 wt %) are added to the mix. The samples are made following the ASTM standards using molds of silicon. The samples are cured at 100°C during 0, 30, 45 and 60 min. Some samples are left without curing treatment for comparison..

For the mechanical characterization, the impact strength tests are performed by using an Izod Impact tester according to ASTM D-256, a specimen size of 5 x 12.7 x 63.5 mm and a Tinius-Olsen Impact Machine at room temperature. The Vickers Hardness test is performed by using a load of 50g, hardness is measured using a Micromet 2003 Vickers Durometer. All of the mechanical property values are obtained by averaging at least five measurements. The fracture surfaces after the impact tests are investigated using a stereo-microscope (Nikon, Japan).

RESULTS AND DISCUSSION

Mechanical properties

Figures 1 and 2 illustrate experimental results obtained from hardness and impact tests measurements. They show the behavior of material with different percentages of starch (5, 10 y 15 wt %). In these figures it can be observed that with increments of the starch content, there is an increase in both properties. It is also shown that the curing time has an important influence on both properties, when it increases the adhesion of particles of starch with the polymeric matrix is improved. This happens because the experiments are within the range corresponding to the gelation temperature of starch producing an increment in these properties. The gelation process begins at curing times below 30 min, longer periods promote formation of a very rigid crystalline regions of starch causing aging of the material. Therefore, the addition of starch helps to reduce the life time of the polymeric matrix and reinforce the mechanical properties of the composite material.

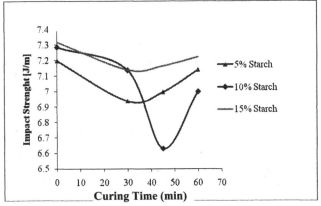

Figure 1. Experimental results obtained from impact tests for samples with 5, 10 and 15 wt % of starch, as a function of curing time.

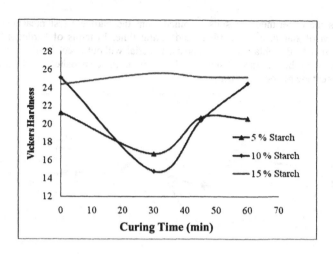

Figure 2. Variation of the Vickers hardness of composite with different content of starch and curing time.

Figure 3 shows experimental results obtained from impact test measurements as a function of starch content, coconut fibers percentage and curing time. In this figure it can be seen that large contents of coconut fibers improve the impact strength. This is most likely due to effect of the fibers to avoid a fast dissipation of the fracture energy. Thus there is an increase in the fracture toughness of the composites due to a good adhesion between the polymeric matrix and the coconut fibers.

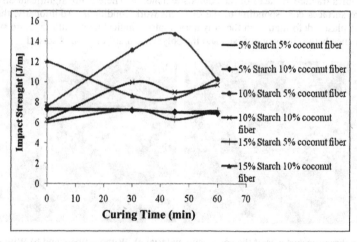

Figure 3. Experimental results obtained from impact test for samples with 5, 10 and 15 wt % of starch, reinforced with 5 and 10 wt % of coconut fibers respectively, at different curing times.

Figure 4 shows the experimental results obtained from the hardness test measurements as a function of % wt of starch, coconut fibers and curing time. In terms of hardness the values obtained are similar to the obtained of composite material without coconut fibers, except by a major variation due to the presence of superficial porosity in the composite material according is increase the percentage of coconut fibers.

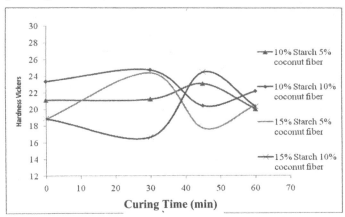

Figure 4. Experimental results obtained from impact test for samples with 5, 10 and 15 wt % of starch, reinforced with 5 and 10 wt % of coconut fibers respectively, at different curing times.

Surface fracture analysis

Figure 5 shows fracture surfaces of the composite material. There is not significant difference in the fracture surfaces corresponding to the different work conditions. Additionally, there is not evidence of plastic deformation in the polymeric matrix (indicating a brittle fracture surface). A good adherence between coconut fibers and the polymeric matrix is also observed

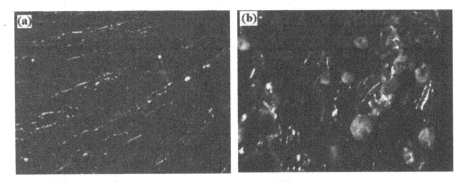

Figure 5. Fracture surfaces of the composite material a) Without fibers and b) With coconut fibers.

CONCLUSIONS

Epoxy resin/starch/coconut fibers composites have been successful fabricated by the described methodology. The mechanical behavior of the polymeric matrix/starch material with coconut fibers under different work conditions is investigated. According to the experimental results it is found that starch helps to decrease life time, and reinforce the polymeric material under certain conditions of temperature. Increments in the amount of coconut fibers give riser to an enhancement of the mechanical properties of the material, this is most likely due to a good adhesion between the polymeric matrix and the fibers.

REFERENCES

1. M. Kolybaba, L.G. Tabil, "Biodegradable Polymers: Past, Present, and Future", ASAE Paper No. RRV03-0007

2. M. José L., K. Herbert, Gil A., Gañán P., "Evaluación de la degradación ambiental de materiales termoplásticos empleados en labores agrícolas en el cultivo de banano en Colombia", Polímeros ciencia e Tecnología, Vol. 17, No. 003, p. 201-205, (2007).

3. M. S. Tomás J., A. V. M. y V. Moreno F., "Potencial de las Fibras Naturales para su uso Industrial", Ciencia Ergo Sum, Toluca Estado de México, (2002).

4. Mutjé P. and Llop M. F, "Desarrollo de Materiales Plásticos Reforzados con Fibras Lignocelulósicas", Iberoamerican Congress on Pulp and Paper Research, (2002).

5. Á. V. B. Celina y V. Analía, "Comportamiento mecánico de compuestos de celulosa modificada/almidón con fibras de sisal cortas", Jornadas SAM/CONAMET/SIMPOSIO MATERIA, (2003).

6. F.-L. Jin, S.-J. Park, "Impact strength improvement of epoxi resins reinforced with a biodegradable polymer", Materials Science& Engineering, A 478, p. 402-405, (2008).

Mater. Res. Soc. Symp. Proc. Vol. 1276 © 2010 Materials Research Society

Magnetic properties of bulk composite FeBSiM (M=Cr,Zr) alloys with high microhardness.

S. Báez[1], I. Betancourt[1], M. E. Hernandez-Rojas[2] and I. A. Figueroa[1].
[1]Departamento de Materiales Metálicos y Cerámicos, Instituto de Investigaciones en Materiales, Universidad Nacional Autónoma de México, México D.F. 04510, México.
[2]Departamento de Biotecnología, Departamento de Ingeniería de Procesos e Hidráulica, Universidad Autónoma Metropolitana-Iztapalapa, México D.F. 09340, México.

ABSTRACT

The synthesis, mechanical and magnetic properties of bulk composite $Fe_{72}B_{19.2}Si_{4.8}M_4$ (M=Cr, Zr) alloys obtained by a copper mould injection casting technique under a protective helium atmosphere are reported and discussed. The resultant microstructure of the composite alloys consists of crystalline $Fe_{92}Si_8$ and Fe_2B phases embedded in a glassy matrix. The values of microhardness (H_v) show maxima for the alloy containing Cr with 10.24 ± 0.95 GPa. The maximum value of saturation magnetization average (μ_0M_s) is 1.25 ± 0.02 T for Cr-containing alloy. The Curie temperatures (T_c) of amorphous phases are higher than 390 K for both alloys. However, the bulk composite alloys presents values for crystallization temperature (T_x) of 1289 ± 10 K and 1140 ± 10 K for Cr and Zr-containing alloys, respectively. These results are interpreted on the basis of the interplay between the crystalline and amorphous phases.

INTRODUCTION

Functional materials are a fundamental part of modern technology due to the increasing necessity of active and versatile solid components performing in a wide range of applications like sensing, actuating, gauging, controlling, monitoring, energy converting, replacing and storing [1, 2]. In particular, the widespread interest on magnetic materials has created an active research field that includes both fundamental investigation and technological applications, ranging from simple toys, to sophisticated imaging techniques devices for the human body [3, 4]. Most modern applications make use of conventional crystalline alloys, which have been studied in detail since the development of the first magnetic steels by the end of the XIX century [5, 6] and whose magnetic properties are determined to a large extent by the careful tailoring of (a) microstructure: phases, grain size and its distribution, grain boundaries, crystal defects, impurities and, (b) chemical composition, which influences intrinsic magnetic properties such as the μ_0M_s (which corresponds to the atom magnetic moment per unit volume of the material; μ_0 is the magnetic permeability of free space), T_c (which indicates the temperature for the ferromagnetic-paramagnetic transition) and magnetocrystalline anisotropy (which refers to existence of easy axis for the magnetic polarization along specific unit cell directions). Less common amorphous alloys, characterized by a lack of atomic long range order, and thus, opposing to crystalline materials, have been subject of research interest since the announcement of the first metallic glass prepared by means of rapid solidification techniques [7]. For magnetic amorphous alloys, the absence of crystallinity constitutes the main factor to produce materials with very high magnetic permeability and readily controllable magnetostriction (i.e. the ability of the alloy to modify its dimensions under the application of an external magnetic field, which strongly depends initially on the transition metals Fe, Co or Ni content). In both types of materials (crystalline and amorphous), the combination with good mechanical properties in general paves the way to more applicable materials with the ability, for instance, of sustaining a magnetic behavior under harsh working conditions such as continuous wear. The objective of the present study is to investigate the effects of small concentration of Cr and Zr on the synthesis, structural, mechanical (hardness) and magnetic properties of the injection cast FeBSiM (M=Cr, Zr) alloys.

EXPERIMENTAL TECHNIQUES

Fe-based alloy ingots of composition $Fe_{72}B_{19.2}Si_{4.8}M_4$ (M=Cr,Zr) are prepared by arc-melting with high purity (99.99 %) elements in a Ti-gettered inert Ar-atmosphere. The alloy compositions represent the nominal values but the processing weight losses are negligible (<0.1 %). The alloy ingots are inverted on the hearth and re-melted several times, in order to ensure compositional homogeneity. Cylindrical alloy rods having a single diameter of 3 mm and a length of ~55 mm are produced by copper mould injection casting under a protective He-atmosphere. The alloys microstructure is characterized by X-ray diffraction (XRD) with Cu-K_α radiation and a Ge primary monochromator (BRUCKER-AXS) is used to reduce the fluorescence of the XRD scan. Scanning Electron Microscopy (SEM) operated at 20 keV and 700 pA and Transmission Electron Microscopy (TEM) operated at 120 keV are also used to determine the structure of the as-cast alloys. The thermal stability of the amorphous phase within each bulk alloy is monitored by using Differential Thermal Analysis (DTA) at a heating rate of 20 K/min. The magnetic properties of the alloys are determined in a Vibrating Sample Magnetometer (VSM) with a maximum applied field of 500 kA/m and a Magnetic Thermogravimetric Analyzer (MTGA) with a heating rate of 10 K/min. The average values and the associated uncertainty are calculated after 10 measurements with fresh samples in every experiment. The H_v of the alloys is carried out using a commercial microhardness tester under a constant indentation load of 50 g during 20 s. The average values and the associated uncertainty are calculated after 20 indentations. The bulk composite alloys are placed in quartz tubes; the air is evacuated to a pressure of 3×10^{-5} kPa, then backfilled with 30 kPa argon and finally sealed. The samples are then heat treated at 1000 K for 8 hours. After this, the samples are removed from the furnace and left to cool down to room temperature.

RESULTS

The presence of the glassy structure and the crystalline phases is confirmed by XRD for both alloys, as shown in Figure 1. In this figure, the existence of diffraction peaks together with a broad maximum between $2\theta=40°$ and $55°$ can be seen. The resulting hump peak of the repeated XRD analysis appeared at ~43.56 for Cr and ~45.49 degrees for Zr. The differences between the hump peaks could be attributed to changes in the composition. In the alloy with Cr (Fig. 1a) the well defined peaks located at 2θ=65.2° and 82.6°, coincide with the (200) and (211) planes of the $Fe_{92}Si_8$ phase. As for the Zr-containing alloy (Fig. 1b), the peaks located at 2θ=79.5° and 92.9°, coincide with the (330) and (332) planes of the Fe_2B phase.

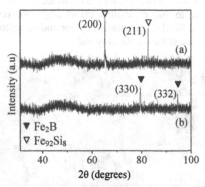

Figure 1. XRD patterns for both alloys: (a) $Fe_{72}B_{19.2}Si_{4.8}Cr_4$ and (b) $Fe_{72}B_{19.2}Si_{4.8}Zr_4$.

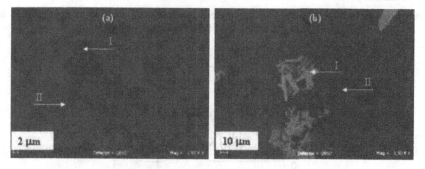

Figure 2. SEM backscattered electron images showing two distinguishable regions (I and II) for both alloys: (a) Cr-containing alloy and (b) Zr-containing alloy.

In addition, SEM backscattered electron images (Fig. 2) show two distinguishable regions (I and II) for the Cr and Zr-containing alloys. Zone I in both alloys (Figs. 2a-b) exhibits a defined structure of equiaxed grains, typical of crystalline phases [8]. Zone I in the Zr-containing alloy can be associated with the Fe_2B phase. In contrast for region II, no crystalline features are evident, and therefore this region is assumed to be the amorphous phase. This region has Fe, Si and Cr contents of ~74 at.%, 7.3 at.% and 2.7 at.%, respectively (according to EDS analysis). Bearing in mind that B cannot be detected in this technique, the measured compositions agree satisfactorily well with expectations for the bulk composite alloys. Similarly, for the Cr-containing alloy, region I can be associated to the $Fe_{92}Si_8$ crystalline phase, and region II to the amorphous phase. This region has an approximate Fe, Si and Zr content of ~75 at.%, 5.8 at.% and 2.9 at.% (EDS analysis).

157

In order to support the aforementioned results, a number of bulk samples are analyzed by TEM. The bright field (BF) images and SAD patterns taken from the cross section of the Zr alloy are shown in Figure 3. The pattern of Fig. 3a is taken from the marked zone in Figure 3b. It consists only of a diffuse halo with no distinctive evidence of crystalline rings (glassy zones). Figure 3c shows an indexed pattern taken from the encircled crystalline zone in Figure 3b. Here several rings corresponding to the crystalline phase Fe_2B are observed. These results are consistent with the XRD and SEM results shown earlier and therefore confirm that a composite structure is successfully produced for the Zr alloy rod.

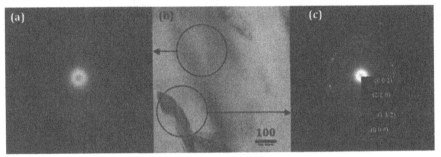

Figure 3. (a) TEM bright field image and (b) electron diffraction pattern taken from region I and (c) crystalline phase of the Zr containing alloy.

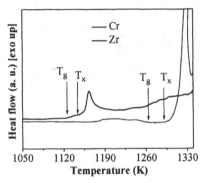

Figure 4. DTA traces for FeBSiM alloys, showing the T_g and T_x: (a) M=Cr and (b) M=Zr.

The DTA results show that the crystallization event for the amorphous phase of Cr-containing alloys occurs at ~1289 K (T_x), with a visible glass transition (T_g) at ~1264 K, as shown in Figure 4a. The Zr-containing alloys show a T_x of 1140 K and a T_g of 1124 K (Fig. 4b). The H_v obtained for both composite bulk alloys are 10.24±0.95 GPa and 9.28±0.17 GPa for the Cr and Zr-containing alloys, respectively. As for the hardness tests, the distance between indentations are ~ 2.5 indenter diameters to avoid interaction between the work-hardened regions of the crystalline zones and the plastic deformation waves of the glassy phase. A summary of physical properties for both composite alloys are shown in Table I. The magnetic properties are investigated with VSM and MTGA. The $\mu_0 M_s$ (within experimental uncertainty) is 1.25±0.02 T for the Cr-containing alloy and 0.97±0.01 T for the Zr-containing alloy (Figs. 5a-b) and the MTGA curves (Figs. 6a-b) show two magnetic transitions for both alloys. These transitions can be associated with different magnetic phases.

Table I. Mechanical, thermal and magnetic properties for the investigated FeBSiM alloys.

M	H_v (GPa)	T_g (K)	T_x (K)	ΔT_x (K)	T_c^{am} (K)	$\mu_0 M_s$ (T)
Cr	10.24±0.95	1264±10	1289±10	25±3	410±5	1.25±0.02
Zr	9.28±0.17	1124±10	1140±10	14±3	400±5	0.97±0.01

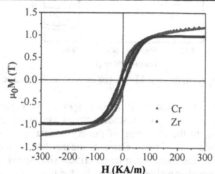

Figure 5. Hysteresis loops for both composite alloys: (a) M=Cr and (b) M=Zr alloys.

For the bulk composite M=Cr alloy (Fig. 6a) two T_c are evident at T_{c1}=410 K and T_{c2}=955 K. The T_{c1} transition can be ascribed to the amorphous phase, since related glassy $Fe_{1-x}B_x$ and Fe-B-Si alloys show a maximum T_c value at ~740 K [9,10]. Based on this, a high temperature regime for T_c values can be defined as T>740 K, in which T_{c2} is contained. These values of high temperatures coincide very well with the T_c of crystalline $Fe_{92}Si_8$ phase (T_c=950 K [9]). Similarly, the bulk composite alloy M=Zr also presents two T_c (shown in Fig. 6b). Again the low temperatures T_{c1}=400 K can be associated to the amorphous phase as is mentioned before, whilst the high temperature T_{c2}=1010 K should be corresponding to Fe_2B phase (with T_c=1015 K [11,12]). Additionally Figure 6c shows a curve obtained from the Cr-containing alloy after heat treatment at 1200 K for 8 hours. It can be seen that the Curie transition corresponding to the amorphous phase disappears after this treatment confirming the existence of this phase in the as-cast alloys. These results agree with the SEM and TEM analysis.

DISCUSSION

The composite nature of the bulk $Fe_{72}B_{19.2}Si_{4.8}M_4$ (M=Cr, Zr) alloys are verified by XRD, SEM, MTGA and TEM results. With these results, it is possible to point out the presence of the crystalline phases Fe_2B and $Fe_{92}Si_8$. It is also found that these phases are dispersed within the amorphous phase, generally having a compositional variance of Fe content. For both composite alloys, the crystallization event of the amorphous phase consists of a single stage. The high T_x observed for the amorphous phase are indeed among the highest T_x reported in, at least, the Fe-based glassy alloy family [13,14]. This enhancement of the resistance to devitrification can be explained by the addition of Cr/Zr atoms with a large atomic radii (r_{Zr}=0.160 nm and r_{Cr}=0.125 nm [15]), relative to the B atom (with r_B=0.082 nm [15]).

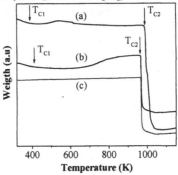

Figure 7. MTGA traces for alloys: (a) as-cast M=Zr, (b) as-cast M=Cr and (c) crystallized M=Cr.

The combined presence of large/small atomic species promotes a more efficient occupancy of interstitial spaces among major constituent atoms, yielding to an improved packing density of the liquid and hence, to the hindering of the crystallization process. Moreover, the high T_x caused by the minor addition of these elements means that the atoms will need more energy to form the critical nuclei, and therefore, a larger critical size would be needed to become a nucleation site. This leads to a higher activation energy for the nucleation and growth processes during crystallization.

The H_v recorded for the M=Zr alloy is found to be higher than some shock-resisting steels (S and M50 type), whose maximum hardness is 7.57 Gpa [16]. For a homogeneous glass structure, it has previously been found that thermal stability, expressed by T_g or T_x, correlates quite well with mechanical properties, such as H_v, at least within a particular alloy system [17]. Nevertheless, the fact that the obtained H_v results are the hardness average of the glassy and crystalline phases restrict comparison but they provide an idea of the bulk hardness of the aforementioned composite alloys. Additionally, the H_v hardness values of both bulk composite alloys can be directly associated with the hardness of the $Fe_{92}Si_8$ and Fe_2B crystalline phases, since this crystalline intermetallic phase is harder and stiffer than the glassy one. The observed μ_0M_s results for Cr-containing alloy are comparable with the saturation polarization of the best low-boron, FINEMET-type alloys ($Fe_{73.5}Si_{13.5}B_9Nb_3Cu_1$), in which the maximum μ_0M_s value of 1.25 T is reported [18]. These excellent μ_0M_s values result from the contribution of the amorphous and crystalline magnetic phases with the latter having μ_0M_s as high as 1.68 T ($Fe_{92}Si_8$ [11]) and 1.40 T (Fe_2B [9]). Similarly, the T_c values observed for the amorphous phases (within the temperature range 400 - 450 K, Table I) of the composite alloys investigated compare very well with the T_c of similar FINEMET-type alloys (between 360 and 843 K [19]). These good values of T_c for the current bulk composite alloys might also be attributed to the considerably large atomic radius of Cr and Zr in comparison to the atomic radius of B. It is thought that the inclusion of such big atoms can produce an enhancement of the integral exchange between magnetic atoms, via an enlargement of the nearest Fe-Fe interatomic distances, and hence, the observed high T_c. Additionally, the increasing values for the T_c of the amorphous phase for both bulk composite alloys can be a consequence of the variable Fe:B ratio observed within the glassy matrix which also affects the first neighbour Fe-Fe separation.

CONCLUSIONS

Bulk composite $Fe_{72}B_{19.2}Si_{4.8}M_4$ (M=Cr, Zr) alloys are successfully produced by copper mould injection casting. The results of T_x are very promising and supersede the range of temperatures at which these alloys are normally exposed. The values of H_v for both alloys are surprisingly higher than some high speed steels. Concerning the magnetic properties, the Cr-containing alloy has a saturation polarization value of 1.25 ± 0.02 T, which is as good as some Fe-Si-B amorphous alloys. The aforementioned properties can be produced by the addition of atoms with a large atomic radii and the constructive interaction between crystalline and amorphous phases.

ACKNOWLEDGEMENTS

The authors are grateful to Adriana Tejeda, Gabriel Lara, Esteban Fregoso-Israel, Omar Novelo, Carlos Florez Morales and Hermilo Zarco for their valuable technical assistance. S. Báez also acknowledges the scholarship granted by CEP, UNAM.

REFERENCES

[1]. Functional Materials, European Conference on Advanced Materials and Processes, Euromat, K. Grassie, E. Tenckhoff, G. Wegner, J. Haubelt and H. Hanselka (Eds), Vol. 13, Wiley-VCH, Weinheim, 2000.
[2]. Materials Science Forum, Y. Umakoshi, S. Fujimoto (Eds), Vol. 512, Trans. Tech. Pub. LTD, Beijing 2006.
[3]. Magnetism, materials and applications, É du Trémolet de Lacheisserie (Ed), Springer, New York, 2005.
[4]. K. H. J. Buschow, Handbook of Magnetic Materials. Vol. 10. K. H. J. Buschow (Ed), Elsevier Science, Holland, 1997.
[5]. B. D. Cullity and C. D. Graham, Introduction to Magnetic Materials, 2nd Ed., IEEE Press Wiley, Hoboken, 2009.
[6]. R. M. Bozorth, Ferromagnetism, Van Nostrand, New York, 1968.
[7]. W. Klement, R. H. Willens and P. Duwez, Nature 187 (1960) 869.
[8]. R.E. Smallman and A.H.W. Ngan, Physical Metallurgy and Advanced Materials, Seven Edition (2007) p.40.
[9]. R. Hasegawa and R. Ray, J. App. Phys. 49 (1978) 4174
[10]. H.S. Chen, Rep. Prog. Phys. 43 (1980) 402.
[11]. M.F. Littmann, IEEE Trans. Magn. 1 (1971) 48.
[12]. L.I. Berger and B.R. Pamplim, Handbook of Chemistry and Physics, CRC, Section 12, Properties of Solids, D.R. Lide (ed) (2005-2006) p.93.
[13] W.H. Wang, C. Dong and C.H. Shek, Mater. Sci. Eng. R 44 (2004) 45.
[14]. Z. Long, H. Wei, Y. Ding, P. Zhang, G. Xie and A. Inoue, J. Alloys Comp. 475 (2009) 207.
[15]. O.N. Senkov and D.B. Miracle, Mater. Res. Bull. 36 (2001) 2183.

[16]. American Society for Metals, Handbook Vol.1, Properties and selection: Irons, Steels and High-Performance Alloys, Wrought Tool Steels, A.M. Bayer, T. Vasco and L.R Walton (Rev.) (1990) p.757.

[17]. I.A. Figueroa, R. Rawal, P. Stewart, P.A. Carroll, H.A. Davies, H. Jones and I. Todd, J. Non-Cryst. Solids 353 (2007), p. 839.

[18]. K. Suzuki, in: Y. Liu, D.J. Sellmyer, D.Shindo (Eds.), Processing and modeling of novel nanocrystalline soft magnetic materials, (Tsinghua University Press-Springer, China, 2006) p.340.

[19]. M.E. McHenry, M.A. Willard and D.E. Laughlin, Prog. Mater. Sci. 44 (1999) 291.

Mater. Res. Soc. Symp. Proc. Vol. 1276 © 2010 Materials Research Society

Microstructure Characterization of Textured Nickel Using Parameters of Extinction

A. Cadena Arenas[1], T. Kryshtab[1], J. Palacios Gómez[1], G. Gómez Gasga[1], A. De Ita de la Torre[2], A. Kryvko[3]

[1]Instituto Politécnico Nacional-ESFM, Av. IPN, Ed. 9, U.P.A.L.M., 07738, México, D. F. México.

[2]Area of Material Science, UAM-Unidad Azcapotzalco, Av. San Pablo #180, 02200, México, D. F., México.

[3]Instituto Politécnico Nacional-ESIME Zacatenco, Av. IPN, Ed. 5, U.P.A.L.M., 07360, México, D.F. México.

E-mail: *camirdena@yahoo.com.mx*, *tkrysh@esfm.ipn.mx*

ABSTRACT

X ray diffraction (XRD) is the common technique for texture analysis by means of pole figure (PF) measurement. PF reflects the grains orientation distribution but contains no information about grain microstructure. The reflected intensity can be affected by the extinction phenomenon that reduces the pole density (PD). The parameters of extinction are related to the crystal microstructure. The parameter of the primary extinction is connected with domain size and parameter of the secondary extinction is related to the angle of domains disorientation that depends on dislocation density in domain boundary. An original XRD method is proposed for correction of PD, considering extinction phenomenon, and separation of the extinction parameters in the case of textured aluminum. The problem is solved under some assumptions. In the present work cold rolled nickel with and without annealing at 600 °C is investigated. The validity of the proposed assumptions for Ni is evaluated in terms of the extinction length. The corrected PD in the maximum of PF and the parameters of the primary and secondary extinction are calculated using the first order reflection for Cu K_α - and Co K_α - radiations and the second order reflection for one of the used wavelengths. Both in cold rolled sample without annealing and in the annealed sample the primary and secondary extinctions are present simultaneously. According to the obtained parameters of extinction the microstructure of textured nickel is evaluated and their modification at the annealing process is demonstrated.

Keywords: X-ray diffraction; texture; extinction; microstructure; dislocations

INTRODUCTION

In polycrystalline materials grain orientations are rarely randomly distributed. The preferred orientations or, more concisely, texture can arise due to anisotropy of a solid [1]. The common quantitative X-ray texture analysis is based on pole figure (PF) measurement by means of X-ray diffraction (XRD) techniques [2]. PF is characterized by pole density (PD) obtained from the reflected intensity of X-rays and does not contain information about microstructure such as size and lattice perfection of crystallites. At PF measurement can take place extinction phenomenon (FP) that reduces PDs, cannot be avoided and must be taken into account [3, 4].

The conventional XRD methods for evaluation of microstructure in polycrystalline materials are based on diffraction peak broadening and the kinematical scattering theory [5]. But they cannot be used in the case when full-width at half maximum of the diffracted peak reaches the instrumental breadth. In real crystals with dislocation densities up to $N_d < 10^8$ cm^{-2}, the X-rays dynamic scattering processes come into being and lead to changes of the reflected intensity [6, 7]. In PF measurements the reflected intensity can be affected not only by PD, but also by EP. The EP originates from two reasons. The primary extinction (PE) takes place in perfect and large domains due to dynamical scattering processes and the secondary extinction (SE) occurs in the crystals with similarly oriented domains by additional diffraction of the incident or the diffracted beam from another domain [8]. Thus, the characteristics of EP are concerned with the crystal microstructural feature. The coefficient of the PE (ε) and of the SE (g) for reflected intensity have been introduced into the XRD kinematical theory [9, 10]. Recently an original XRD technique has been proposed to eliminate the influence of EP in texture analysis [11]. The application of this technique allows correcting PD and obtaining parameters of the PE and SE. Since the problem cannot be solved exactly, some assumptions are proposed, suitable in the case of aluminum for which the wavelength of the K-absorption edge is distant from commonly used wavelengths. Nickel has the wavelength of the K-absorption edge very close to these wavelengths [12].

We evaluated the validation of the approximations [11] for Ni samples taken into account the variation of the dispersion corrections for the atomic scattering amplitude owing to close position of K-absorption edge with respect to wavelength using. The microstructure determination of cold rolled nickel with and without annealing at 600 °C is performed by the use of the primary and secondary extinction parameters obtained at the application of an original technique [11]. The microstructure modification at the annealing process is demonstrated.

EXPERIMENTAL DETAILS

Nickel samples after 75 % cold rolling with and without annealing are used for PF measurements by XRD technique. Nickel samples after cold rolling are heated in a furnace at the heating rate of 10 ^{0}C/min to the temperature of 600 ^{0}C and held at this temperature during 30 min. A nickel powder standard sample is also measured at the same conditions as textured samples. For texture analysis one-axis D8 Advance Bruker X-ray diffractometer with an Euler cradle and with two non-polarized Cu and Co radiations is used. Pole figures for <001> and <111> crystallographic directions are measured for the first and second order reflections. The measurement duration for weak second order reflections is several times longer as compared with the duration for low index reflections.

RESULTS AND DISCUSSION

The presence of texture in the samples studied is determined by using of XRD patterns measured with Cu K_α - radiation and simulated one. The changes in the peak intensities distribution in XRD patterns for the samples with respect to the distribution in simulated one for powder are observed for both samples that confirmed the texture formation.

The direct pole figures obtained for 111 and 222 reflection for nickel samples after cold rolling and with annealing at 600 °C are showed in Figure 1 and Figure 2, respectively.

The comparison of the PFs shows that PF for 222 reflection has more details. Additionally, PDs in the maximum for the second order reflection is higher than PDs in the maximum for the first order reflection for both samples that indicates the present of EP. So, for texture analysis of these samples an original XRD technique [11] can be used after evaluation of assumptions validation.

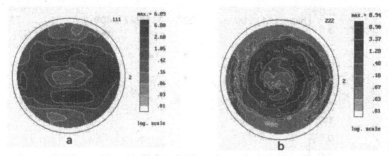

Figure 1. Direct pole figures for nickel after 75 % cold rolling obtained with Cu-Kα radiation for 111 reflection - (a) and for 222 - reflection - (b).

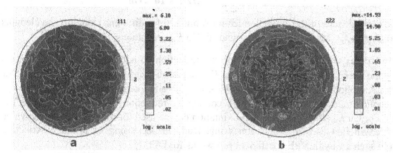

Figure 2. Direct pole figures for nickel after 75 % cold rolling and annealing at 600 °C obtained with Cu-Kα radiation for 111 reflection - (a) and 222 reflection - (b).

The first assumption is that for low index reflections the PE coefficient *ε* does not depend on wavelength used and has the average value for different wavelength. The parameter of PE can be evaluated within the terms of domain thickness *l* and extinction length Λ [13]:

$$\varepsilon = \tanh(l/\Lambda)(l/\Lambda)^{-1}. \qquad (1)$$

For the symmetrical Bragg case and non-polarized radiation Λ can be presented as [11]:

$$\Lambda = \frac{1}{2d} \frac{v_c}{|C_{dyn}| r_0 F_H}, \qquad (2)$$

where *v* is the unit cell volume, r_0 is the electron orbit radius, *d* is the spacing of the lattice-planes (*hkl*), $C_{dyn} = (1 + \cos 2\theta)/2$ is the polarization factor for dynamical scattering, F_H is the structure factor that depends on the atomic scattering amplitude and on the dispersion corrections, which are determined by the ratio of the K- absorption edge wavelengths for Ni (1.49 Å) and the wavelength of radiation used. The dependences of Λ on reflections are shown in Figure 3.

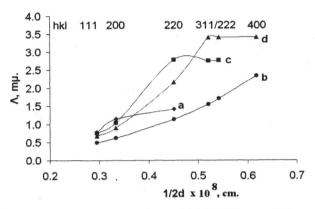

Figure 3. The dependences of the extinction length Λ on 1/2d for different radiation wavelengths for nickel: (a) Cr K_α -,(b) Mo K_α -, (c)Co K_α - and (d) Cu K_α - radiations.

The extinction lengths for 111 reflection and for 200 reflection do not change essentially for Co K_α -, Cu K_α - and Cr K_α - radiations, but for Mo K_α –radiation the difference is more than 25%. Thus the first assumption is valid for Ni for Co K_α -, Cu K_α - and Cr K_α– radiations.

The second assumption is that for the second order reflection the PE coefficient (ε) is equal to unity. For Cu K_α - radiation this approximation can be used for nickel, because the value of extinction length increases in a few times that leads to decreasing of primary extinction phenomenon in such a way that PE coefficient (ε) works for unit.

Hence, for nickel samples it is possible to use the proposed assumptions using Co K_α - , Cu K_α - radiations for measurements and the system of three equations as proposed in [11]:

$$P_{m(\lambda 1)}^{hkl}\left(\mu_1 + g\varepsilon Q_{kin1}\, C_2/C_1^2\right) = P_{cor}\mu_1\varepsilon ,$$
$$P_{m(\lambda 2)}^{hkl}\left(\mu_2 + g\varepsilon Q_{kin2}\, C_2/C_1^2\right) = P_{cor}\mu_2\varepsilon , \qquad (3)$$
$$P_{m(\lambda 2)}^{2(hkl)}\left(\mu_2 + g Q_{kin3}\, C_2/C_1^2\right) = P_{cor}\mu_2 .$$

In this system a set of two equations for the same first order reflection, measured with different wavelengths, and one equation for the second order reflection are used.

Table I. Calculate values of the absorption coefficient μ (cm^{-1}) and the extra term of $Q_{kin}C_2/C_1^2$ (cm^{-1}) for Co K_α -, Cu K_α -, Cr K_α - and Mo K_α - radiations for different reflections.

Radiation	μ	$Q_{kin}C_2/C_1^2$ (111)	$Q_{kin}C_2/C_1^2$ (222)	$Q_{kin}C_2/C_1^2$ (200)	$Q_{kin}C_2/C_1^2$ (400)
Co Kα	628.15	1.42	0.47	1.23	---
Cu Kα	407.18	1.03	0.45	0.84	0.27
Cr Kα	1283.04	4.34	---	4.2	--
Mo Kα	415.206	0.27	0.06	0.21	0.04

The influence of the SE on the reflected intensity can be evaluated by comparing the ratio between the values of absorption coefficient μ and the term $Q_{kin}C_2/C_1^2$ for different wavelengths and reflection orders. The calculated values of these parameters are presented in Table I.

The directly measured PDs in maxima for different orders of reflection, crystallographic directions and wavelengths used, as well as the corrected PDs and calculated parameters of extinction by the proposed method [11] are presented in Table II. The values of PDs for low-index reflections are smaller than those for high order reflections with both radiations due to strong influence of the EP on reflected intensity. The directly calculated PDs in maximum for 111 reflection show that the annealing leads to decreasing of the texture for <111> direction. But for 222 reflection, where the influence of EP on reflected intensity decreases essentially, the values of PD increased for both radiations used. The influence of EP on PD for 200 reflection is strong and EP decreased the PD with respect to 400 reflection in five times. So, the use only the first order reflections for evaluation of texture modification at the annealing results in incorrect interpretation of the annealing process. The differences in PDs for the first and second order reflections for different crystallographic directions varied. These variations connected with different influence of EP on reflected intensities owing to microstructure anisotropy.

Table II. Calculated parameters for Ni after cold rolling (R) and with annealing at 600°C (A). $P_{m(Co)}$, $P_{m(Cu)}$- measured PD, P_{corr}-corrected PD, ε- PE coefficient, g – SE coefficient, l -average size of domain, ū - average angle of disorientation of domains.

<hkl>	$P^{<hkl>}_{m(Cu)}$	$P^{<hkl>}_{m(Co)}$	$P^{2<hkl>}_{m(Cu)}$	P_{corr}	ε	g	l μm	ū min	$N_D \times 10^{-7}$ (cm^{-2})
<111>R	6.89	6.9	8.94	8.98	0.99	180.84	0.67	3.79	17.6
<111>A	6.10	6.20	14.93	16.65	0.43	161.44	1.54	4.20	8.56
<200>R	4.70	4.80	9.10	12.61	0.71	433.00	0.99	1.58	4.96
<200>A	13.30	12.70	62.00	76.88	0.22	661.32	1.80	1.03	1.78

From parameter of the PE ε and ratio of l/Λ [12] the coherent-domain thickness l is calculated. Average angle of domains disorientation ū is calculated from the value of g, according to ū = $(2g\sqrt{2\pi})^{-1}$. It is also possible to evaluate the dislocation density N_D in domain boundaries within the framework of the mosaic crystal model using the calculated values of l and ū as [14]:

$$N_D = \frac{\sqrt{2}}{3\sqrt{\pi}} \frac{\bar{u}}{lb} , \qquad (5)$$

where b is the magnitude of Burgers vector. For Ni the value of Burgers vector as $b = (a/2)[110]$ (a is a lattice parameter) is used. The results of the calculations are presented in Table II. The average sizes of domains after cold rolling for both <111> and <001> directions are less than the ones after additional annealing. During the annealing the size of coherent perfect domains increases due to the gliding motion of dislocations to domains boundaries and the average angle of domains disorientation decreases due to the annihilation of dislocations in these boundaries. This process is more intensive for (111) plane, which is the glide plane for the f.c.c. structure. Apparently, the dislocation density in domain boundaries is somewhat overestimated because the size of coherent perfect domain is not the real size of the domain if some amount of dislocations is present in it. For microstructure evaluation we used the model of the mosaic crystal that does not take into account this situation. The complete theory for the estimation of the dislocation density at their nonuniform distribution does not exist. But in any case, the obtained information about the microstructure can be used for explanation of the annealing process not only as changing of texture, but also as a dislocation structure modification in grains.

CONCLUSIONS

An evaluation has been made for Ni regarding the validity of the proposed assumptions in original XRD technique for the analysis of textured materials. This is done by considering that the extinction phenomenon is present in the terms of the extinction length. The data obtained show that the extinction phenomenon is present in textured cold rolled nickel and after thermal annealing. PDs in direct pole figures are underestimated by different values for different directions. The application of this technique for PDs correction due to the elimination of the extinction effect is extremely useful in the PD determination for the correct interpretation of the thermal annealing process. The determined parameters of the primary and secondary extinction for textured Ni are used for microstructure evaluation. The average coherent perfect domain size, the average angle of domains disorientation and the dislocation density in the domain boundaries are calculated. The obtained values are different for different directions due to the anisotropic inelastic deformation properties of nickel. The modification of domain size and dislocation structure is demonstrated.

ACKNOWLEDGEMENTS

The authors would like to thank the CONACyT of Mexico for financial support of this work by the projects N 83425 and N83781.

REFERENCES

1. Kocks U. F., Tomé C. N. and Wenk H.R.. Texture and Anisotropy, Cambridge University Press (1998).
2. Randle, V. and O. Engler. Introduction to Texture Analysis Macrotexture, Microtexture and Orientation Mapping, Gordon and Beach Science Publisher, Amsterdam (2000).
3. Mucklich, A., Klimanek, P. *Mater. Sci. Forum.* **157-162**, 275 (1994).
4. Yamakov, V., Tomov I. *J. Appl. Cryst.* **32**, 300 (1999).
5. M.A. Krivoglaz. *X-Ray and Neutron Scattering in Nonideal Crystals* (Springer Verlag, Berlin – Heidelberg - New York, 1996).
6. A. Authier, F. Balibar, and Y. Epelboin, *Phys. Stat. Sol.* **41**, 225 (1970).
7. V.L. Indenbom and V.M. Kaganer, *Phys. Stat. Sol. (a)* **87**, 253 (1985).
8. R.W. James. *The Optical Principles of the Diffraction of X-Rays* (3rd ed. London: G.Bell, 664 (1965).
9. W.H. Zachariasen, *Acta Cryst.* **16**, 1139 (1963).
10. W.H. Zachariasen, *Acta Cryst.* **23**, 558 (1967).
11. T. Kryshtab, J. Palacios-Gómez, M. Mazin, and G. Gómez-Gasga. *Acta Materialia* **52/10**, 3027 (2004).
12. International Tables for X-ray Crystallography (Dordrecht, Boston, London: Kluwer Acad. Publ., 1992).
13. Z.G. Pinsker, *Dynamical Scattering of X-Rays in Crystals* (Springer-Verlag, Berlin-Heidelberg - New York, 1978).
14. I.A. Larson and C.L. Corey, *J. Appl. Phys.* **40** 2708 (1969).

Mater. Res. Soc. Symp. Proc. Vol. 1276 © 2010 Materials Research Society

Mechanofusion Processing of Metal-Oxide Composite Powders for Plasma Spraying

Ricardo Cuenca-Alvarez[1], Carmen Monterrubio-Badillo[2], Hélêne Ageorges[3], Pierre Fauchais[3]
[1] Instituto Politécnico Nacional, CIITEC, México, D.F. 02250, Mexico
[2] Instituto Politécnico Nacional, CMP+L, México DF 07430, Mexico.
[3] SPCTS-UMR 6638, University of Limoges, 87060, Limoges Cedex, France.

ABSTRACT

Composite particles destined to build plasma sprayed coatings, are prepared by the mechanofusion process (MF). These particles consist of a stainless steel core particle coated by finer particles of alumina. Changes induced by the MF process are monitored by SEM, DRX, and laser granulometry, revealing that the dry particle coating process is governed by agglomeration and rolling phenomena. Simultaneously, the MF performance is controlled by the operating parameters such as the compression gap, the mass ratio of host to guest particle, and the powder input rate. The mechanical energy input leads to a nearly rounded shape of the final composite particles; however, no formation of new phases or components decomposition is detected by XRD analysis. The resulting composite powder features optimal characteristics, concerning particle shape and phases distribution, to be plasma sprayed in air.

INTRODUCTION

Production of composite powders is performed from a variety of commercial processes; however powder characteristics depend on the parameters of each process. Examples are ball milling, sol-gel process, atomization, mechanical alloying, sintering, self-propagating high-temperature synthesis (SHS), spray drying, etc. [1]. Since applications such as plasma spraying, need to use particles featuring characteristics such as good flow ability, spherical shape, two or more phases joined with no reaction between each other; powder preparation route must be selected according to these needs. Then, it is not only advantageous, but also necessary to design devices with the ability to modify such materials and thus obtaining composite powders with unique functionality. Some devices are developed to achieve this goal [2]: The Hybridizer©, the Magnetically Assisted Impaction Coater (MAIC), the Rotating Fluidized Bed Coater (RFBC), the Theta Composer, and the Mechanofusion Process©. These techniques are based under the dry particle coating principle with no need for using binder or water [3]. Consequently, economical and environmental benefits are obtained because some stages are avoided, for instance, drying of powder. A special interest is oriented towards the Mechanofusion device, henceforth called MF. It can be used for inducing significant changes in the functionality or properties of the original host particles, and thus creating engineered powders [4]. This technique attaches fine particles (guest) onto the surface of much larger particles (host), via strong mechanical forces; simultaneously, spheroïdization of particles is attained by a rolling effect.

Mechanofused powders have been used for different applications. For instance, organic particles type PMMA (polymetyl-metacrylate) coated by a film made of TiO_2 increase their flowability. Similar results are found when resin particles are coated by a carbon film [5]. The MF process is also used to prepare iron particles coated by ceramic glass for magnetic or electronic applications; spherical particles of synthetic graphite in battery production; deagglomerated and spherical coloring particles for printer inkjet; composite particles made of nickel and aluminium for applications of light metals at high temperature [7]. Some fuel cells components are prepared by the MF process for instance based on nickel-cobalt oxide helping to decrease nickel dissolution when melting of carbonate occurs [8]. Thus, in the following sections it will be described the influence of main parameters of processing, firstly on deformation of metallic particles and, secondly the particle coating. The powder system is chosen by considering a review of the bibliography oriented towards a wear resistance application [9-15].

EXPERIMENTAL PROCEDURE
Materials and the Mechanofusion Process

Stainless steel (SS) is specified as the host particles, whereas alumina (AL) as the guest ones. The last is sustained by the increase of wear resistance by combining toughness of metals with hardness of ceramics. Physical characteristics and SEM micrographs of raw powders are given in table I. Since dry particle coating depends on the particle size distributions (PSD) of host and guest particles, materials are chosen with a very large difference in PSD as confirmed by laser granulometry. Commercial gas atomized 316L stainless steel is provided by Sultzer Metco with two particle size distributions whereas finer α-alumina is from Baikowzki, France.

Table I. Powder characteristics

Function	Host	Guest
Material	316L Stainless Steel	Alumina (α- phase)
Mean Particle size [μm]	142	1.5
Specific mass [kgm⁻³]	7960	3900
Designation	SS	AL
Morphology		

Preparation of composite particles is performed by using an in-house designed MF set-up, described previously [16]. It consists of a cylindrical chamber rotating on the vertical axis at 1400 rpm, with a concentric joint of compression hammers and scraper blades remaining static. The gap between the inner wall of the chamber and the compression hammer is adjustable. Due to centrifugal forces and, depending on the compression gap, the powder is forced against the chamber wall and dynamically compressed through the gap. Consequently, particles bed is intensively mixed and subjected to different phenomena such as compression, attrition, frictional shearing or rolling. Then, mechanical energy input, plus the generated heat can lead to mechanical alloying, homogeneous mixing, or deformation of metallic particles.

When two different types of particles, in terms of chemical composition and particle size distribution, are MF processed; the finest particles (secondary) are attached on the coarser particle surface (host) without needing to use binders [5]. There are several operating parameters affecting the performance of the MF device [6]. However, once the characteristics of host and guest particles are specified, the key parameters are the rotation speed, processing time as a

170

function of the powder input rate, compression gap and the mass ratio of host to guest particles. Then, a compression rate (τ) is defined by the relation between the powder bed thickness formed over the inner wall (EC) and the spacing of compression gap (EF):

$$\tau = \frac{EC - EF}{EC} \quad ....(1)$$

As mentioned above, this work analyzes firstly the influence of compression rate affecting the stainless steel particle shape, followed by the study of feasibility of the MF device to coat stainless steel host particles by pure alumina in function of the powder charges and the powder rate input. The corresponding variations in the operating parameters are given in table II.

Table II. MF operating parameters.

Parameter	Value
Stainless steel charge [g]	150
Compression rate [τ]	25, 15, 5
Mass ratio of Host/Guest particle	15, 7.5, 3
Processing time [h]	1, 2, 3, 4, 5

Sample characterization

Changes induced by the MF device on the processed particles are followed by X-ray diffraction (XRD, Cu-Kα radiation, Siemens, model D-5000), scanning electron microscopy (SEM, Phillips, model XL30), energy dispersive spectroscopy (EDS, Oxford, Link ISIS), and laser granulometry (Malvern, Mastersizer Hidro2000). To analyze the cross sections of particles by SEM, powder samples are embedded in a Mecaprex resin. Then, surfaces for analysis are prepared by standard metallographic procedure. Polished surfaces are coated with a thin layer of gold (45 nm in thickness) to avoid polarization onto ceramic and resin areas.

RESULTS AND DISCUSSION
Deformation of metallic particles

Compression rate plays an important role on the particle shape. When $\tau = 25$, metallic particles are welded to the internal chamber wall due to the overheating generated by the friction between the compression hammer and the powder bed (figure 1a). However, friction decreases rapidly at lower values of τ (15) where deformed particles are obtained as shown in figure 1b. For τ=5, the compression gap is widely spaced to induce a moderate deformation of particles with a tendency to spheroidize them (figure 1c). Nevertheless, fine particles (~1 µm) appear as a result of the abrasion effect taking place into the wide gap formed by the geometry of scraper blades. Thus, in order to reduce this effect, scraper blades geometry is modified to recover more efficiently the agglomerated powder from the wall surface with an incipient abrasion effect.

Coating of stainless steel particles by Al$_2$O$_3$

By using $\tau = 5$, milling and overheating of particles is avoided but a rolling effect is still present, particle coating is strongly influenced by the behavior of alumina particles into the powder bed. When alumina content is evaluated by means of processing powder at values of mass ratio of host to guest particles of 3 and 7.5, alumina particles are segregated onto the chamber wall surface as shown in Fig. 2a. This behavior allows just the coverage of some particles featuring a heterogeneous surface coating (Fig. 2b). However, if MF process is performed with smaller amounts of fine guest particles, surface coverage is more uniform. This phenomenon applies for a mass ratio of host/guest particles of 15.0 and, is explained by a better dispersing of alumina particles within the bed of particles, avoiding their segregation.

Figure 1. Shape of mechanofused particles at different τ: (a) 25. (b) 15 and (c) 5.

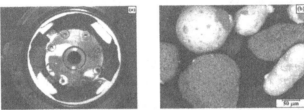

Figure 2. (a) Agglomeration of alumina particles onto chamber components. (b) SEM micrographs of mechanofused particles at higher alumina contents.

By considering the latest, processing time is investigated as a function of the powder input rate by introducing alumina particles at 0.05 g/min in order to ensure a well dispersion of both phases into the powdered bed. Samples are taken by intervals of 1 h up to 5 h. A comparison in particle size distributions of mixtures processed at different periods (Fig. 3a), shows a slight difference in the main peak centered at 105 μm. It is likely that, even though the compression gap is widely spaced, a strong rolling effect is still induced, thus attrition of coarser metallic particles take place in the early stages of MF processing, reducing the size of metallic particles. However, a small peak is observed in the range of 0.3 to 1 μm for the samples processed up to 4 and 5 h. This phenomenon suggests that guest alumina particles, which previously have been attached to the surface of host particles, now are detached due to their successive passing throughout the compression gap. In Fig. 3b, the corresponding XRD patterns reveal an increase in size of Al_2O_3-α peak as more alumina particles are introduced. Nevertheless, a slight oxidation of stainless steel particles is detected on 47° 2θ in all cases as a result of attrition taking place in the early stages of processing described before. Then, material not oxidized is renewed at the surface of metallic particles, but oxidation does not continue because of attaching of alumina particles onto that metallic surface preventing its wearing. Attrition and deformation effects, described above, lead the composite particles to adopt a spherical shape after 4 h of processing, achieving a shape factor of 1.25 (1.0 corresponds to the perfect sphere). Morphology and cross-sections from the resulting composite particles (fig. 4) reveal the formation of a uniform coating of alumina onto the surface of stainless steel particles, attaining up to 5.4 μm of thickness.

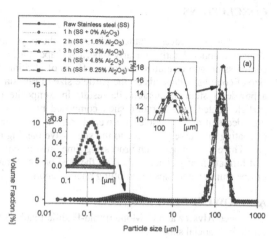

Figure 3. Evolution of (a) PSD and. (b) XRD patterns during MF process of stainless steel SS plus alumina at different processing times: 1, 2, 3, 4, and 5 h.

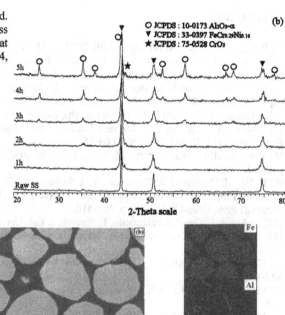

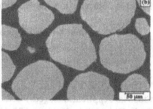

Figure 4 (a) Morphology and (b) cross section of stainless steel/Al₂O₃ mechanofused powders together with EDS mappings for iron (Fe) and aluminum (Al) in grey areas.

CONCLUSIONS

The MF process changes effectively the physical properties of individual particles in a targeted manner. Firstly, a nearly spherical shape is achieved by using a sufficiently wide spaced compression gap to induce moderate effects of attrition or deformation. Consequently, the final size of composite depends on the particle size of raw stainless steel. Later, the MF processing of a powder system of metal/oxide results in composite particles consisting of a stainless steel particle coated by a ceramic shell composed by Al_2O_3. This particle coating is homogenous provided that guest particles are well dispersed within the particle bed. Nevertheless, detaching of coating may take place due to the successive passing of particles throughout the compression gap. Therefore, an optimum processing time must be considered up to 5 h. Finally, it is proposed that formation of a ceramic shell on the host particles surface is governed by the agglomeration of fine alumina particles via a rolling phenomenon.

ACKNOWLEDGEMENTS

R. Cuenca-Alvarez and C. Monterrubio-Badillo thank CONACYT-Mexico and SFERE-France for their financial support.

REFERENCES

[1] R. Davies, Powder Technology, (119), 2001, p. 45-57.
[2] R. Pfeffer, R.N. Dave, D.Wei, M. Ramlakhan, Powder Technology, (117), 2001, p. 40-67.
[3] B.H. Kaye, Powder Mixing 1st ed., (London, England, Chapman & Hall, 1997), p. 132-147.
[4] M. Alonso, M. Satoh, K. Miyanami, Powder Technology, (59), 1989, p. 45-52.
[5] T. Yokoyama, K. Urayama, M. Naito, M. Kato, KONA, (5), 1987, p. 59-68.
[6] R. Cuenca Alvarez, Ph.D Thesis (in French), University of Limoges, France, Nov. 2003.
[7] J. Stein, M. Tenhover, T. Yokoyama, Ceramic Forum International / Ber.DKG, Process Engineering, vol. 79, (4), 2002, p. E11-E15.
[8] T. Fukui, H. Okawa, T. Hotta, M. Naito, T. Yokoyama, J. Amer. Cer. Soc., vol. 84, (1), 2001, p. 233-235.
[9] H. Ito, M. Umakoshi, R. Nakamura, T. Yokoyama, K. Urayama, M. Kato, in Thermal Spray Coatings: Properties, Processes and Applications, (ed.) T.F. Bernecki, (pub.) ASM International Materials Park, Ohio-USA, 1991, p. 405-410.
[10] G; Farnè, F. Genel Ricciardiello, L. Kucich, J. Eur. Cer. Soc., vol. 19, 1999, p. 347-353.
[11] A. Csanády, A. Csordás-Pintér, L. Varga, L. Tóth, G. Vincze, Mikrochim. Acta, (125), 1997, p. 53-62.
[12] M. Egashira, T. Kato, T. Hyodo, Y. Shimizu, Key Engineering Materials, vol. 247, 2003, p. 427-432.
[13] H. Herman, Z.J. Chen, C.C. Huang, R. Cohen, J. Thermal Spray Tech., (12), 1992, p.129-135.
[14] D. Bernard, O. Yokota, A. Grimaud, P. Fauchais, S. Usmani, Z.J. Chen, C.C. Berndt, H. Herman, in Thermal Spray: Industrial Applications, (ed.) C.C. Berndt and S. Sampath, (pub.) ASM International Materials Park, Ohio-USA, 1994, p. 171-178.
[15] H. Ageorges, P. Fauchais, , Thin Solid Films, (370), 2000, p. 213-222.
[16] R. Cuenca-Alvarez, H. Ageorges, P. Fauchais, P. Fournier, A, Smith, Materials Science Forum, Vol. 442, (2003), pp. 67/72.

Mater. Res. Soc. Symp. Proc. Vol. 1276 © 2010 Materials Research Society

Stress concentration on artificial pitting holes and fatigue life for aluminum alloy 6061-T6, undergoing rotating bending fatigue tests

Víctor H. M. Lemus, Gonzalo M. D. Almaraz and J. Jesús V. Lopez
Universidad Michoacana (UMSNH), Santiago Tapia No. 403, Morelia Michoacán, 58000, México.

ABSTRACT

This work deals with rotating bending fatigue tests on aluminum alloy 6061-T6, under loading condition close to the elastic limit of the material. Results have been obtained for three types of specimens: without artificial pitting, specimens with one artificial pitting hole and specimens with two neighboring artificial pitting holes. Results show that fatigue endurance is reduced in the case of one pitting hole and considerably for two neighboring pitting holes. In order to explain this behavior, numerical analysis by FE are carried out to determine the stress concentrations for the three types of specimens. It is found that the stress concentration for two neighboring pitting holes is an exponential function of the separation between the two holes, under uniaxial loading. The probability to find two or more neighboring pitting holes in real industrial materials, such as cast iron, corroded or pitting metallic alloys is high; then, the stress concentration for two or more neighboring pitting holes needs to be considered for the fatigue prediction life under fatigue loading and corrosion attack applications.

INTRODUCTION

The aluminum alloy 6061-T6 is a precipitation hardening alloy with high content of magnesium and silicon, presenting good mechanical properties and weldability. It is one of the most common aluminum alloys for general purpose use: aircraft fittings, brake pistons, hydraulic pistons, appliance fittings, valves and valve parts, bike frames, camera lens mounts, couplings, marines fittings and hardware, electrical fittings and connectors, decorative or misc. hardware, hinge pins, magneto parts and others.

Modern industrial applications of aluminum alloys imply frequently environmental corrosion attack; this is the reason of some recent works dealing with the problem: development of the "Corrosion Pit Growth Law" and corrosion fatigue lives [1], surface corrosion protection of aluminum structures on marine environments, non destructive quantification of pittings, pitting corrosion behavior of aluminum alloy on welded joint, the effect of temper conditions and corrosion on the fatigue endurance of an aluminum alloy, improving pitting corrosion resistance of aluminum alloy by laser surface melting, or proposing a probability model for the growth of corrosion pits in aluminum alloys.

This work is devoted to the study of fatigue endurance of aluminum alloy 6061-T6 under rotating bending fatigue tests, when one or two artificial pitting holes are machined at the narrow section of the hourglass shape specimen. Special attention is focused on the stress concentration factors caused by the artificial pitting holes and the relationship to experimental fatigue endurance.

EXPERIMENTAL PROCEDURE

Tests are carried out at room temperature. A cooling air system is implemented in order to keep the testing temperature below 60 °C at the critical specimen narrow section. Under this condition, no modification of the crystallographic structure in the testing material is expected. The machining process for all specimens is as homogeneous as possible in order to avoid important variation on the surface roughness. The average value for SRz, the "maximum height roughness" is 14 μm. In Figure 2 are shown the artificial pitting holes machined on the narrow section of specimens: Figure 2a for one single artificial pitting hole and Figure 2b for two neighboring artificial pitting holes.

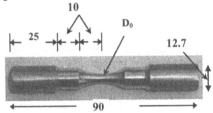

Figure 1. Specimen shape and dimensions (mm).

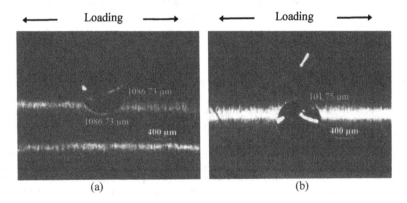

Figure 2. Artificial pitting holes: (a) single pitting hole, (b) Two neighboring pitting holes.

The diameter of artificial pitting holes Dp is comprised between 1060 and 1140 μm. The separation of the two pitting holes is about 100 μm; this last parameter is controlled by automatic machining. Uniaxial fatigue loading is achieved for the three types of testing specimens; the stress concentration factor for the single pitting hole (hemispherical surface cavity), is evaluated according the expression [2]:

$$Kt = 1.522\left(1 + \frac{2}{7 - 5v}\right) \qquad (1)$$

Here v is the Poisson coefficient (for an aluminum alloy $v = 0.33$). The stress intensity factor Kt is 2.09. There is an important increase of stress is at the bottom of hemispherical cavity for one artificial pitting hole [3]. The stress is higher for the same applied load P in the case of two neighboring pitting holes located at the narrow section of specimen. The stress concentration factor and fatigue endurance relationship for tested specimens is analysed in further sections of this work.

In order to determine the loading condition and the stress distribution inside the specimen, a numerical simulation by means of the software Visual Nastran (specimens without pitting) and Ansys are carried out, see Figure 3. It is found that with a bending load of $P = 39$ N, the induced Von Mises stress at the narrow section of the specimens without pitting is close to σ_n =105 MPa, the 39 % of the elastic limit of this material (270 MPa).

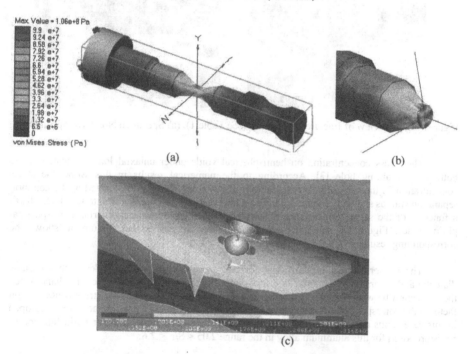

(a)   (b)

(c)

Figure 3. Numerical simulation for (a) a specimen without pitting hole, (b) clipping at the narrow section, and c) stress distribution for specimen with two neighboring artificial pitting holes.

The pit aspect ratio a/2c (the depth of pit "a" and the diameter "2c") is a main parameter for stress concentration [3]. These results agree with analytical results presented in the middle of last century [4]. Nevertheless, two neighboring pitting holes under uniaxial loading can induce stress

177

concentration as a consequence of their proximity when the geometrical dimensions of pitting holes remain constant. In this work, an exponential function for the relationship between the stress concentration factor K_t and the proximity of two pitting holes under uniaxial loading is presented.

RESULTS AND DISCUSSION

Figure 4 shows the lateral view of fractured samples with one and two artificial pitting holes. In both cases the fracture follows the diameter direction perpendicular to the applied load and passes through the hemispherical pitting holes.

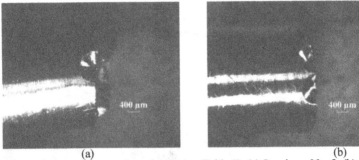

(a)   (b)

Figure 4. Lateral view of fractured specimens (see Table 1). (a) Specimen No. 9, (b) Specimen No. 24.

The stress concentration on hemispherical voids under uniaxial load is located at the bottom of a pitting hole [3]. According to the numerical results in this work, the stress concentration in the case of two neighboring pitting holes seems to be located at the common separation wall as shown in Figure 3c. In order to investigate the stress concentration factor Kt as a function of the separation distance S between the centers of two neighboring hemispherical pitting holes, Figure 5a, some numerical simulation are carried out. Figure 5b shows the corresponding results.

The numerical results show that for two hemispherical pitting holes with the same diameter and separated by S/r = 2.2, the stress intensity factor is Kt = 3. On the other hand, when the separation between the two hemispherical pitting holes is S/r = 2.6, the stress concentration factor is the corresponding for a single hemispherical pitting hole i.e., no interaction is developed in this case. Thus an empirical formulation for the stress concentration factor Kt in function of S/r is proposed for this aluminum alloy in the range 2.05 < S/r < 2.6:

$$Kt = 1.76 \ (S/r - 2)^{-0.33} \qquad (2)$$

Table 1. Experimental parameters and results.

Test No.	Without Pitting	One Pitting	Two Close Pitting	Test Frequency f (Hz)	Apply Load (N)	Kt-σn/σy (%)	No. of Cycles (Fatigue life)
1	√			50	39	39	291000
2	√			50	39	39	240500
3	√			50	39	39	210000
4	√			50	39	39	299500
5	√			75	39	39	231600
6	√			50	39	39	355500
7	√			50	39	39	228000
8	√			25	39	39	204500
9		√		50	39	80	89450
10		√		50	39	80	89550
11		√		50	39	80	62000
12		√		50	39	80	96000
13		√		50	39	80	57000
14		√		50	39	80	53700
15		√		50	39	80	124750
16		√		100	39	80	119400
17		√		100	39	80	109000
18		√		100	39	80	71500
19		√		50	39	80	93250
20		√		25	39	80	102000
21			√	25	39	113	35475
22			√	25	39	113	25750
23			√	50	39	113	32400
24			√	50	39	113	26300
25			√	50	39	113	30650
26			√	25	39	113	26850
27			√	25	39	113	25375
28			√	25	39	113	31520

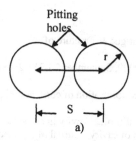

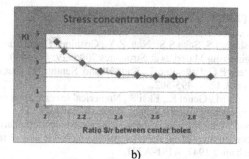

Figure 5. Stress concentration factor: a) geometrical parameters between two neighboring pitting holes, b) evolution of stress concentration factor with the ratio S/r between center holes.

The "Multiaxial fatigue limit criterion for defective materials" has been presented in recent works [5] relating fatigue endurance and stress distribution. It is corroborated that the gradient of the hydrostatic part of the stress distribution at the tip of the defect is a good parameter to represent the defect influence on the fatigue resistance in the range of the fatigue limit under multiaxial loading. Nevertheless, the analytical development based on one single defect, which is a very hypothetical case in real materials with surface defects (cast iron, corroded or pitted material), cannot represent the real hydrostatic stress gradient at the interacting zone of two neighboring hemispherical holes. Further investigation is necessary for the understanding of fatigue-corrosion phenomena; particularly, the interaction between fatigue crack growth and pitting growth [6]; pitting holes geometrical dimensions and proximity; grain boundaries, size, shape, and orientation versus fatigue-corrosion [7]; multiaxial loading and others factors.

CONCLUSIONS

Fatigue endurance under rotating bending fatigue tests of aluminum alloy 6061-T6 decreases with the presence of one artificial pitting hole and dramatically with two neighboring artificial pitting holes. For industrial applications of metallic alloys enduring corrosion attack and fatigue, the probability to generate two or more neighboring pitting holes is high; then, fatigue-corrosion design should consider the presence of high concentration stresses induced by two or more neighboring pitting holes. Stress concentration for two neighboring pitting holes seems to be located at the common separation wall. The range of test frequency: 25 to 100 Hz, does not affect fatigue life of three types specimens, as it is shown on Table 1. The stress concentration factor increases exponentially with the proximity of two neighboring pitting holes; an exponential expression is presented in this work for a range of proximity. For real materials with pitting holes (cast iron, corroded or pitting elements), the probability to find two or more neighboring pitting holes is high; then, stress gradients should be estimated under this condition, not for a single pitting hole.

ACKNOWLEDGMENTS

The authors acknowledge the Universidad de Michoacan (UMSNH, Morelia, Mexico) for the development of this work. Special mention of gratitude to CONACYT (National Counsel for Science and Technology, Mexico City) for the financial support devoted to this study.

REFERENCES

1. Ishihara S, Saka S.S., Nan Z.Y., Goshima T. and Sunada S., Fatigue & Fracture of Engineering Materials & Structures 2006; 29: 472–480.
2. Paris P.C, Palin-Luc T., Tada H. and Santier N., Crack Paths 2009, September 23-25th 2009, Vicenza, Italy: 495-502.
3. Cerit M., Genel K., Eksi S., Numerical investigation on stress concentration of corrosion pit, Engineering Failure Analysis 2009; 16: 2467–2472.
4. Sadowsky M.A., Sternberg E., Stress concentration around an ellipsoidal cavity in an body under arbitrary plane stress perpendicular to the axis of revolution of cavity, Journal of Applied Mechanics 1947, A191-A201.
5. Nadot Y., Billaudeau T., Multiaxial Fatigue Limit Criterion for Defective Materials, Engineering Fracture Mechanics 2006; 73:112-133.
6. Van der Walde K., Hillberry B.M., Characterization of pitting damage and prediction of remaining fatigue life, International Journal of Fatigue 2008; 30:106–118.
7. Jones K., Hoeppner D.W., The interaction between pitting corrosion, grain boundaries, and constituent particles during corrosion fatigue of 7075-T6 aluminum alloy, International Journal of Fatigue 2009; 31: 686–692.

Mater. Res. Soc. Symp. Proc. Vol. 1276 © 2010 Materials Research Society

Synthesis of La₄Ni₃O₁₀ Cathode Material (SOFC) by SOL-GEL Process

$$\text{Synthesis of La}_4\text{Ni}_3\text{O}_{10} \text{ Cathode Material (SOFC) by SOL-GEL Process}$$

Rene Fabian Cienfuegos[1,2], Sugeheidy Carranza[1], Leonardo Chávez[1,2], Laurie Jouanin[3], Guillaume Marie[4], Moisés Hinojosa[1,2]

[1]Fac. Ing. Mec. Eléc. (FIME), San Nicolás de los Garza, Nuevo León, 66451, México
[2]Cent. Innov., Invest. Desar. Ing. Tec. (CIIDIT), Apodaca, Nuevo León, 66600, México
[3] Université Paris-Sud 11, Orsay, 91405, France.
[4] IUT A de Lille, Boul. Paul LANGEVIN, BP 179, Villeneuve d'Ascq Cedex, 59653, France

ABSTRACT

The goal in this study is to synthesize a Ruddleden-Poper La-Ni phase (La₄Ni₃O₁₀) using a polymeric route. This material exhibits mixed ionic and electronic conduction (MIEC) properties and can be used as cathode material in the manufacture of Solid Oxide Fuel Cells (SOFC). . In addition, an easy and inexpensive synthesis method is presented The polymeric precursors are prepared following the Castillo method using optimized the complexation ratios (HMTA/metallic salts) from 1 to 6. The obtained powders are characterized by differential scanning calorimetry (DSC), thermogravimetric analysis (TGA) and X-ray diffraction (XRD) in order to determine the processing conditions for formation of the crystalline phase. Experiments performed using complexation ratios of 5 and 6 do not show coagulation. However, the solution prepared using a complexation ratio of 5, is transformed into a gel after few days. Gels produced from solutions prepared with complexation ratios from 2 to 5 were heated at 800, 900 and 1000°C to obtain solid materials. These powders are characterized by TGS, DSC and XRD and it is found that the temperature needed to obtain crystalline La₄Ni₃O₁₀ was 1000°C.

INTRODUCTION

Electricity production has used fossil fuels for many years. Recently the price of these fuels has increased considerably as a result of decreasing worldwide reserves . In addition, abuse in the use of fossil fuels increases greenhouse gas emissions (GHG) which contribute to global warming and increase pollution with negative impact on the environment. Experts in the field of energy technology agree that the use of fuel cells for power generation can substantially reduce oil dependency and its negative environmental impact. Fuel cell devices convert chemical energy directly into electrical energy without combustion as an intermediary step. This increases conversion efficiency when compared to conventional power generation combustion systems with the added benefit of lower emissions of gases like CO_2 , NO_x and SO_x [1-6].

Fuel cells are classified according to the type of electrolyte used in their manufacture.alkaline fuel cells (AFC) use an an alkaline electrolyte and operate at temperatures between 70 and 100 °C.Molten carbonate fuel cells (MCFC) utilize lithium and molten potassium as electrolyte And Phosphoric acid fuel cell (PAFC) operate at temperatures between 180 and 210 °C while proton exchange membrane fuel cells (PEMFC) operate at temperatures between 80 y 200 °C.Solid oxide fuel cells (SOFC) are often fabricated using an electrolyte composed of zirconium oxide (ZrO_2) doped with 8% yttrium oxide (Y_3O_3) and can operate at temperatures between 700 y 1000 °C.

Principle of a SOFC

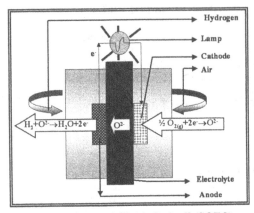

Figure 1 Diagram of basic fuel cell (SOFC)

In this study the interest is on solid oxide fuel cells because they can produce electricity through pre-reforming and can operate with oil or natural gas producing lowergreenhouse gas emissions. They have longer operational life due to the higher stability of the solid electrolyte. . The cell consists of two electrodes (cathode and anode) separated by the electrolyte. The cathode contains the oxidizer and the anode the fuel. The oxygen in the cathode is reduced to O^{2-} anions using the electrons produced in the anode (fig.1).

The development of new energy conversion devices such as SOFC cells, requires cathode materials capable to operate at temperatures between 600 to 1000 °C. In this scheme, the cathode should be constructed using compounds that exhibit both electronic and ionic conduction properties (MIEC, Mixed Ionic and Electronic Conductor) because the oxygen reduction process takes place at the triple point of contact (TPB –Triple Phase Boundaries). Previous work [2] has shown that $La_4Ni_3O_{10}$ can synthesized using a complexation ratio of 3. In this work, the synthesis of $La_4Ni_3O_{10}$ using a polymeric route with complexation ratios between 1 and 6 is optimized.

EXPERIMENTAL METHODOLOGY

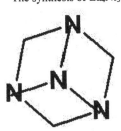

Figure 2 Chemical structure of hexamethylnetetramine

The synthesis of $La_4Ni_3O_{10}$ is carried out by reacting a metallic salt solution with a solution containing the complexing agent. The metallic salt solution is prepared by dissolving lanthanum nitrate ($La(NO_3)_3 \cdot 6H_2O$) and nickel nitrate ($Ni(NO_3)_2 \cdot 6H_2O$) in 20 ml of deionized water. The molar ratio between La and Ni salts in the solution is nLa/nNi=4/3. The complexing agent solution is prepared using hexamethylenetetramine (HMTA), acetylacetone (ACAC) and acetic acid. HMTA has a chemical structure of nitrogen atoms rigidly secured by a complex ring system in a pyramidal configuration (fig. 2). This compound crystallizes into a rhombic dodecahedron structure and can be carbonized at 280 °C. It is soluble in methanol, ethanol and aqueous solutions and is readily hydrolysable to aminomethyl compounds that can be used instead of ammonia as a catalyst for the synthesis of resins. The catalytic action exhibits a a weak state atalkaline pH in the range from 7-10. HMTA has a tendency to explode when present in powder form, while in solutions it is relatively stable and little hydrolysis takes place in the presence of acids. This primary reaction with acids produces formation of salts such as (CH_2) $6N_4$-HCl, $[(CH_2)$ $6N_4]_2 \cdot H_2SO_4$ and $(CH_2)N_4$-H_3PO_4 with low stability which cannot be easily isolated. Additional compounds are produced with chlorine, bromine, iodine and various metal salts.

Acetylacetone (ACAC, 2,4-pentadiona) (fig. 3.), is a colorless or slightly yellow liquid with a characteristic odor and a boiling temperature of 140.6 °C. It is mainly used as a complexation agent in the production of metal oxides such as titanium oxides.

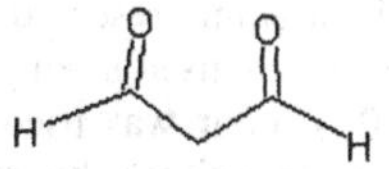

Figure 3 Chemical structure of acetylacetona

Acetic or carboxylic acid (systematic name ethanoic acid), is a colorless liquid with a characteristic odor, boiling point 118°C and pH=2.4 (glacial). This acid is soluble in water, alcohol, glycerine and ether. It is used in the chemical industry as acidulant and a neutralizing agent, for the production of acetic anhydride, acetate monomer and chloroacetic acidfor and in the production of plastics, pharmaceuticals, insecticides, photographic chemicals, food additives and coagulants.

The complexing agent solutions areprepared by mixing the metallic salt solution with a 1:1 equi-molar HMTA/ACAC solution dissolved in 100 ml of acetic acid. Various complexing agent solutions are prepared by varying the stoichiometric ratio $R=n_{HMTA}/(n_{Ni}+n_{La})$ from 1 to 6. The reaction is performed at 60 °C during 15 minutes.

The final product was heated to dry the gel and then calcinatedto obtain the crystalline oxide. Finally, the oxide powders were characterized using TGA ((TGA- Q50 Instrument TA)to follow the lost mass as a function of temperature, DSC (DSC Q10 Instrument TA) to determine the occurrence of any exothermic and/or endothermic reactions and X-ray diffraction (X-ray diffractometer Bruker model D8 Advanced) to identify the different phases formed after heat treatment.

RESULTS AND DISCUSSION

Precursor characterization

The gels are dried at temperatures between 200 and 250 °C and the resulting powders characterized by the TGA and DSC. TGA and DSC curves exhibit a mass loss of about 10 % between 20 a 90 °C, which corresponds to water evaporation. Between 300 and 500 °C the mass loss is about 80% and this process is accompanied by two successive exothermic reactions. Mass spectrometry analysis has shown that the observed exothermic peaks are associated with emission of nitrogen (NO, NO_2) and carbon oxides (CO, CO_2) during decomposition of organic oxycarbonates detected by infrared spectroscopy [3].

Importance of temperature on the material crystallization

Thermal analysis (fig. 4) shows that a significant exothermic reaction occurs just before reaching 400 °C and another publication for the same material (index 3) shows its end at 600 °C [7], the reason for this being is the preheating to 400 °C which allowed the temperature to be stable for 5 hours, This was done in order to eliminate the organic matter and avoid the projection of amorphous powder in the furnace. Another publication shows the existence of a descarboxylation between 700 and 800 °C [3, 7]. Therefore it can be assumed that the presence of crystalline powder material begins at 800 °C [3]. In figure 5 we observe crystallization at 1000 °C, while for 800 y 900 °C is still in its amorphous state (as an example evolution of the diffractogram presented for index 2, respect with temperature). This may be because in the cited papers they used a temperature ratio of 100 °C/hour until it reach 1000 °C while in this report 1000 °C/hour was used, it might not allow thermal inertia at temperatures below a 1000 °C despite to wait two hours for crystallization.

Figure 4 DSC and TGA sample R 2.

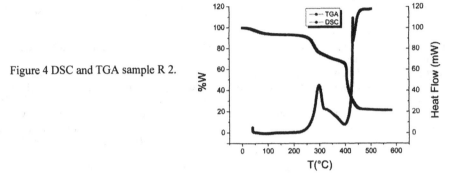

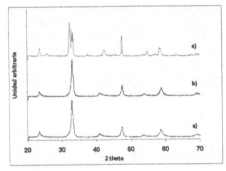

Figure 5 X-ray diffraction to $La_4Ni_3O_{10}$ phase for index 2 at temperature (a) 800, (b) 900 , (c)1000.

Effect of complexing agent index variation of metal salts

In order to evaluate the variation between complexing agents in the structure of the oxides, ratios from 1 to 6 were used. The experiment with index 6 showed coagulation and this is the reason why this diffractogram is not shown, in the case of index 5 (R=5), the solution exhibited coagulation after a week. X-ray diffraction shows the phase obtained with the index 2 to 4, it shows that metal cations complexation is not homogeneous for low organic content (R=5). The case of index 1 shows the La_2O_2 phase, this for low complexing agent quantity. The reason for this, could be the low heat release during the decomposition which did not allowed for the cations homogeneity. Index 5 showed the highest concentration of complexing agents which caused coagulation.

CONCLUSIONS

La$_4$Ni$_3$O$_{10}$ phase was synthesized using the polymeric sol-gel route. Several tests were conducted, varying complexing agent and metallic salt indexes from 1 to 6. The characterization of the material was made using thermal analysis and X-ray diffraction. This proved that using a working temperature of 1000 °C (1000 °C/hour) the La$_4$Ni$_3$O$_{10}$ phase can be obtained for indexes 2, 3 and 4.

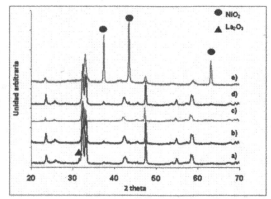

Figure 6 X-ray diffraction at 1000 °C to La$_4$Ni$_3$O$_{10}$ with the ratio (a) 1, (b) 2, (c) 3, (d) 4, (e) 5

REFERENCES

1. F. Mauvy , C. Lalanne, J.M. Bassat, J.C. Grenier, H. Zhao, P. Dordor and Ph. Stevens, Journal of the European Ceramic Society 25, 2669-2672 (2005)
2. S. Castillo, R.F. Cienfuegos, M.L. Fontaine, P. Lenormand, P. Bacchin and F. Ansart, Materials Research Bulletin 42, 2125-2131 (2007)
3. Marie-Laure Fontaine, Christel Laberty-Robert, Antoine Barnabé, Florence Ansart and Philippe Tailhades, Ceramics International 30, 2087–2098 (2004)
4. S. Nesaraj, Journal of Scientific & Industrial Research, 69, 169-176 (2010)
5. K.C. Wincewicz and Joyce S. Cooper, Journal of Power Sources, 140, 280–296 (2005)
6. Chunwen Sun, Rob Hui and Justin Roller, J Solid State Electrochem, 14, 1125-1144 (2010)
7. R. F. CIENFUEGOS, D. Thesis, Université Toulouse III, 2008,

Mater. Res. Soc. Symp. Proc. Vol. 1276 © 2010 Materials Research Society

Alumina-Based Functional Materials Hardened with Al or Ti and Al-nitride or Ti-nitride Dispersions

José G. Miranda-Hernández[1], Elizabeth Refugio-Garcia[1], Elizabeth Garfias-García[1] and Enrique Rocha-Rangel[2]
[1]Departamento de Materiales, Universidad Autónoma Metropolitana Av. San Pablo No. 180, Col. Reynosa-Tamaulipas, México, D. F., 02200.
[2]Universidad Politécnica de Victoria, Avenida Nuevas Tecnologías 5902, Parque Científico y Tecnológico de Tamaulipas, Ciudad Victoria, Tamaulipas, 87137, México

ABSTRACT

The synthesis of Al_2O_3-based functional materials having 10 vol. % of fine aluminum or titanium and aluminum-disperse or titanium-dispersed nitride hardened-particles has been explored. Two experimental steps have been set for the synthesis; specifically, sintering of Al_2O_3-aluminum or Al_2O_3-titanium powders which were thoroughly mixed under high energy ball-milling, pressureless-sintered at 1400°C during 1 h in argon atmosphere and then for the second step it was induced formation of aluminum nitride or titanium nitride at 500°C during different times (24, 72 and 120 h) by a nitriding process via immersion in ammoniac salts. SEM analyses of the microstructures obtained in nitride bodies were performed in order to know the effect of the ammoniac salts used as nitrating on the microstructure of aluminum or titanium for each studied functional material. It was observed that an aluminum nitride or titanium nitride layer growth from the surface into the bulk and reaches different depth as the nitriding time of the functional material was increased. The use of aluminum or titanium significantly enhanced density level and hardness of the functional materials.

INTRODUCTION

Functionally Graded Materials (FGMs) are a new generation of composite materials which allow a more efficient use of existing homogeneous materials (composites) by introducing gradual gradients of composition, microstructure, texture, phase distribution or particle content in their structure [1-2]. Mainly the FGMs have hard phase in the surface and lower hard phase in the body [3]. The first industrial applications of the FGM concept have emerged in production of cermets cutting tools in 1996 by Sumitomo Electric Industries, Ltd [4]. Since then, the FGMs are also applied in where operating conditions are rigorous, for example, rocket heat shields, heat exchanger tubes, thermoelectric generators, heat-engine components, thermomechanical loads, biomaterials, electronic materials and electrically insulating metal/ceramic joints [5]. The FGMs can be constituted by carbide ceramics, oxide ceramics and metals, and they can be used in application at high temperatures such as in the construction of gas turbine engines in order to increase their thermal cycle efficiency. There are several methods in order to produce them, such as: chemical vapour deposition CVD, phisycal vapour deposition PVD, plasma spraying, powder metallurgy and selfpropagating high-temperature synthesis (SHS) [6-8]. It is well known that Al_2O_3-based ceramics possess excellent physical and chemical properties, as well as, good mechanical resistance and thermal stability, but Al_2O_3 ceramics are very sensitive to minimal defects in their microstructure, which acts as point of beginning of cracks. By this disadvantage is necessary the incorporation of fine particles in their matrix of ductile metals, this cause changes and improvement the physical and mechanical properties giving a new material class classified as cermets [6, 9]. Likewise, because of the incentive to improve productivity due to the global competition much effort is being made in FGMs research to find novel techniques and material formulations to increase the performance of such materials. In this work through the use of powders techniques, firstly they will be obtained Al_2O_3-10 vol. % Al and Al_2O_3-10 vol. % Ti cermets and later they will be submitted to a nitriding processes during different times by means of immersion in ammoniac salts with the objective to attain $Al_2O_3/Al/AlN$ or $Al_2O_3/Ti/TiN$ FGMs.

EXPERIMENTAL

Experimental route consists of two steps (cermets fabrication and FGMs fabrication). Cermets fabrication: The raw material in this step were Al_2O_3 powders (99.9 %, 1 μm, Sigma, USA), aluminum powders and titanium powders (99.9 % purity, 1-2 μm, Aldrich, USA). The nominal composition of the final composite materials was Al_2O_3-10 vol. % Al and Al_2O_3-10 vol. % Ti. Original powders were mixed under high energy ball-milling. Then milled powders were used to produce compacted cylinders (200 MPa) with 25 mm diameter and 3 mm thickness. Compacted samples were pressureless-sintered at 1400°C during 1 h in argon atmosphere. The heating and cooling rates were kept constant and equal to 10 °Cmin^{-1}. The characterization of the synthesized products includes the evaluation of density by Arquimedes' method and microhardness measurements with the help of a Vickers indenter. The microstructure of the cermets was investigated by optical microscopy (OM). During the second step it was induced formation of aluminum or titanium nitride by a thermal treatment of cermets previously fabricated by their immersion in an ammonia salts bath at 500°C during different times (24, 72 and 120 hrs). The effect of the ammonia salts was analyzed by scanning electron microscopy (SEM). The SEM was equipped with an energy dispersive X-Rays detector (EDX). Microhardness was also measured in transversal sections of these nitride samples (in all cases ten independent measurements per value were carried out).

RESULTS AND DISCUSSION

Cermets materials

Table I shows values of density, shrinkage and microhardness of the cermets produced. In this table it can be observed that final density of fabricated cermets is important, since relative densities were 95% and 96% for Al_2O_3-10 vol. % Al and Al_2O_3-10 vol. % Ti cermets respectively. Such situation appears to be verified by the strong shrinkage of the samples after they were sintered. Microhardness of Al_2O_3-10 vol. % Ti sample is bigger than that of Al_2O_3-10 vol. % Al sample. Accordingly this behavior is understandable if it is considered that titanium has a hexagonal structure whereas the structure of aluminum is face cubic center, in were the first one is harder than the other and in combination with Al_2O_3 results in a harder material.

Table I. Density values and micro-hardness measured in cermets.

Cermets	Density (g/cm^3)	Relative density (%)	Shrinkage (%)	Micro-hardness (HV)
Al_2O_3-10 vol. % Al	3.68	95	8	62.3 ± 7.1
Al_2O_3-10 vol. % Ti	3.89	96	41	183.8 ± 14.4

Microstructure

Microstructure observed by OM corresponding to Al_2O_3-10 vol. % Al or Al_2O_3-10 vol. % Ti cermets is showed in Figures 1a and 1b respectively, in both cases metallic particles are the brightness phase homogeneously distributed in the opaque phase that corresponds to the ceramic matrix. In both figures a homogeneous distribution of metallic particles can be seen around the ceramic grains. Typical microstructure of cermets that involves a metallic network surrounding the ceramic material, apparently here is not observed principally in figure 1a, or if it exist is too fine to be resolved by light microscopy. This suggests that wetting in the system Al_2O_3-Al was too poor. Therefore the angle of contact during sintering among liquid Al and Al_2O_3 was rather large and higher than 90° and as a consequence Al distribution in the alumina matrix becomes rather uneven [10]. In figure 1b that corresponds to the composite containing Ti, it is observed a fine metallic partial network distributed around some grains of the ceramic matrix. This can be achieved by the high specific surface energies reached by titanium during milling inducing a good distribution of this metal in the solid state, since no liquid phases are produced during the sintering.

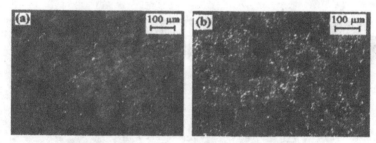

Figure 1. Optical microstructures of sintered cermets.
(a) Al_2O_3-10 vol. % Al and (b) Al_2O_3-10 vol. % Ti.

Functionally Graded Materials

Cermets obtained in the former stage were submitted to a nitriding processes by means of a thermal treatment in an ammonia salts bath at 500°C during different times (24, 72 and 120 h) in order to accomplish the formation of aluminum or titanium nitride. Likewise, the chemical gradient is formed by the nitriding of the metallic particles that are localized close up to the outside of the material. At 500°C dissociation of ammonia molecules takes place and nitrogen is released to diffuse and react with the metallic phase localized at the surface of the cermets forming metallic nitrides (AlN or TiN) in agreement with reactions (1) and (2).

$$Al + N \rightarrow AlN ; \quad \Delta G^{\circ} = -287 \frac{KJ}{mol} \quad (1)$$

$$Ti + N \rightarrow TiN ; \quad \Delta G^{\circ} = -308 \frac{KJ}{mol} \quad (2)$$

Microstructure

Figure 2 shows SEM pictures of the surface cross section of different samples after nitriding treatment. For these observations samples were cut perpendicularly to the compaction direction to investigate nitrogen distribution as a function of depth. Figs. 2a, 2b and 2c, respectively, present the microstructure that resulted after nitriding the Al_2O_3-10 vol. % Al cermets and Figs. 2d and 2e, present the microstructure that resulted after nitriding the Al_2O_3-10 vol. % Ti cermets. In figures that corresponds to Al_2O_3-10 vol. % Al-cermets it can be seen the growth of the surface layer displaying a different color with respect to the cermets since the surface to the core of the sample. This color change phases are due to the formation of metallic nitrides. The depth of the layer was of 12.7, 16.2 and 49.4 µm for Al_2O_3-10 vol. % Al-cermets nitride during 24, 72 and 120 h respectively. In the same way, it is observed a similar behavior in figs. 2d and 2e that correspond to Al_2O_3-10 vol. % Ti-cermet nitride during 24 and 72 h and which present a depth layer of 37 and 44.8 µm respectively. It is necessary to mention that cermets containing Ti collapsed mechanically as a result of multiple cracks after nitriding process for long times 120 h. Most likely the synthesis of FGMs based on Al_2O_3-Ti-TiN is not feasible by this method for long times.

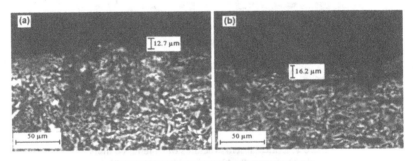

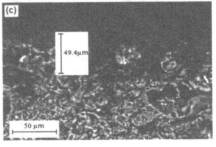

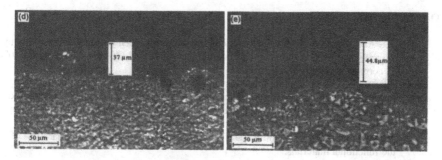

Figure 2. SEM pictures of the surface cross section of samples after nitriding treatment.
(a), (b) and (c) Al_2O_3-10 vol. % Al cermets nitride during 24, 72 and 120 h respectively.
(d) and (e) Al_2O_3-10 vol. % Ti cermets nitride during 24 and 72 h respectively.

Such contrast, in practice exhibited similar texture to the nitride region. So that between the nitride layer and the non-nitride region, there is an intermediate zone which consists of partially-nitride metal particles. Though for analysis EDX it could not be determine the presence of nitrogen for confirm the formation of metallic nitrides, for this reason is appealed to microhardness measurements in order that indirectly decides the presence of nitrides.

Microhardness

Table II shows results of hardness measurements in nitride samples. In this table it is has for Al_2O_3-10 vol. % Al and Al_2O_3-10 vol. % Ti cermets microhardness evaluations. This way starting 10 μm from the edge towards the core of the sample hardness increases as the nitriding treatment time is enlarged. To the extent that the measurement is done at a greater distance (30 μm) from the edge toward the center hardness tends to decrease sharply, however, hardness continues rising with treatment time. When the hardness measurement is performed at 100 μm from the shore, hardness of the samples is similar to the hardness of these after the nitriding treatment. Hereby there is had that the local region in which Al-particles or Ti-particles have reacted with diffusing nitrogen is defined to as the nitride zone, whose thickness turns into layer depending on the metal reinforcement and nitriding time. In order that between the nitride layer and the non-nitride region, there is an intermediate zone which consists of partially-nitride metal particles. Therefore, moving from the outermost surface part into the bulk of material, it has been detected three specific regions, featuring: (1) fully nitride metal particles, (2) partially nitride particles and (3) metallic particles not being nitride.

Table II. Hardness measurements in different parts of samples nitride at different times.

Material	Depth of hardness measurement in sample (μm)	Hardness (HV)			
		0 h	24 h	72 h	120 h
Al_2O_3-10% Al	10	62.3 ± 7.1	247.2 ± 16	316.9 ± 13	566 ± 15
	30		111.5 ± 4	149.4 ± 12	311.5 ± 8
	100		71.8 ± 15	81.2 ± 3	58.8 ± 11
Al_2O_3-10% Ti	10	181.8 ± 6.4	625.8 ± 14	688.9 ± 15	----
	30		433.2 ± 18	483.3 ± 15	----
	100		172.4 ± 15	256 ± 15	----

CONCLUSIONS

o Al$_2$O$_3$/Al/AlN and Al$_2$O$_3$/Ti/TiN functional materials has been prepared successfully by a nitriding process via immersion in ammoniac salts of Al$_2$O$_3$-10% Al and Al$_2$O$_3$-10% Ti cermets previously fabricated.

o Aluminum nitride or titanium nitride layer growth from the surface into the bulk and reaches different depth as the nitriding time of the functional material was increased.

o The use of titanium lead to obtain harder materials with nitride layer of about 44.8 μm in less time compared with the materials prepared with aluminum.

o The use of aluminum or titanium significantly enhanced density level and hardness of the functional materials.

ACKNOWLEDGMENTS

EGG and ERR wish to express their thanks to the SNI for the distinction of their membership and the stipend received. Authors wish to thank the Departamento de Materiales at UAM-A for the financial support given through project 2260235.

REFERENCES

1. Weiping Liu and J.N. DuPont, "Fabrication of functionally graded TiC/Ti composites by Laser Engineered Net Shaping", Scripta Materialia **48** (2003) 1337.
2. A. *Mortensen and S. Suresh,* "Functionally graded metals and metal-ceramic composites", International Materials Reviews **40** (1995) 239.
3. José G. Miranda-Hernández, Elizabeth Refugio-Garcia and Enrique Rocha-Rangel, "Synthesis of Functional Materials by Means of Nitriding in Salts of Al$_2$O$_3$–Al, Al$_2$O$_3$–Ti and Al$_2$O$_3$–Fe Cermets" Materials Research Society Proceedings of The Structural Symposium, IMRC2009, in press.
4. L. Jaworska, M. Rozmus, B. Królicka, A. Twardowska, "Functionally graded cermets" Journal of Achievements in Materials and Manufacturing Engineering, **17** (2006) 138.
5. J. K. Wessel, "The Handbook of Advanced Materials", John Wiley & Sons, New York, (2004).
6. E. Rocha-Rangel and J. G. Miranda-Hernández, "Alumina-Copper Composites with High Fracture Toughness and Low Electrical Resistance", Materials Science Forum **644** (2010) 43.
7. Y. Miyamoto, W. A. Kaysser, B. H. Rabin, A. Kawasaki, and R. G. Ford, (eds.), "Functionally Graded Materials; Design, Processing and Applications", Kluwer Academic, USA, (1999).
8. Walter Lengauer, Klaus Dreyer, "Functionally graded hardmetals", Journal of Alloys and Compounds **338** (2002) 194.
9. S.J. Ko, K.H. Min, Y.D. Kim and I-H: "A study on the fabrication of Al2O3/Cu nanocomposite and its mechanical Properties", Journal of Ceramic Processing Research. **3** (2002) 192.
10. E. Rocha, P. F. Becher and E. Lara, Rev. Soc. Quím., Méx. **48** (2004) 146.

AUTHOR INDEX

Acevedo, J., 85
Ageorges, Hélêne, 169
Almaraz, Gonzalo M.D., 175
Altamirano-Torres, Alejandro, 129, 149
Amézaga-Madrid, P., 111
Amirjanov, Adil, 135
Aparicio, R., 91
Avdeev, Ilya, 135

Báez, S., 153
Ballesteros, C., 31
Barrera, G., 91
Benachour, M., 55
Benachour, N., 55
Benguediab, M., 55
Betancourt, I., 155

Cadena Arenas, A., 163
Calderon, H.A., 97
Carranza, Sugeheidy, 181
Carreño-Gallardo, C., 103
Chávez, Leonardo, 181
Cienfuegos, Rene Fabian, 181
Covarrubias, Octavio, 49
Cuenca-Alvarez, Ricardo, 169

das Neves, Maurício D.M., 61
de Araújo, C. José, 31
De Ita de la Torre, A., 163
de la Garza-Gutiérrez, H., 39
de la Torre, S.D., 39
Delijaicov, Sérgio, 61

Estrada-Guel, I., 103, 111

Fals, Andres E., 21
Fauchais, Pierre, 169
Figueroa, I.A., 155
Filho, Francisco A., 61
Flores, Ismael, 141

García-Vázquez, F., 85
Garfias-García, Elizabeth, 187
Garibay Febles, V., 97
Garza, A., 85
Gómez Gasga, G., 163
Gonzalez, C., 91
Gutiérrez-Castañeda, E., 67
Guzmán-Flores, I., 85

Hadjoui, A., 55
Hastert, Andrew, 135
Hernandez-Rojas, M.E., 155
Herrera-Ramírez, J.M., 103, 111
Hinojosa, Moisés, 181

Jouanin, Laurie, 181

Kryshtab, T., 163
Kryvko, A., 163

Lemus, Víctor H.M., 175
Lima, Emmanuel P.R., 61
López, H.F., 1, 73, 117, 123
Lopez, J. Jesús V., 175
López-Cortés, V.H., 73
López-Cuellar, E., 31, 141
López-Pavón, L., 31

Marie, Guillaume, 181
Martínez-Sánchez, R., 103, 111
Mendoza-Del-Angel, H., 123
Miki-Yoshida, M., 111
Miranda-Hernández, José G., 187
Monterrubio-Badillo, Carmen, 169

Palacios Gómez, J., 163
Pérez-Bustamante, F., 111
Pérez-Bustamante, R., 103, 111
Perez-Medina, G.Y., 73
Plascencia-Barrera, G., 39

Quintero, Julio, 21

Ramírez-Argáez, M.A., 11
Refugio-Garcia, Elizabeth, 187
Reyes-Valdés, F.A., 73
Robles Hernández, F.C., 21
Rocha-Rangel, Enrique, 129, 149, 187

Salazar-Nieto, J., 129
Salinas B., J., 79
Salinas-Rodríguez, A., 67, 79
Sanchez, R., 117
Sandoval-Pérez, Francisco, 129, 149

Santana García, I.I., 97
Sobolev, Konstantin, 135, 141
Solórzano-López, J., 11

Torres, Leticia M., 141
Torres-Castro, A., 31
Torres-Hernández, Yaret G., 149
Trapaga, G., 91

Valdez, Pedro L., 141

Zambrano, P., 73
Zarazua, Elvira, 141
Zenit, R., 11

SUBJECT INDEX

air jet, 11
Al alloys, 55, 91, 111, 175
alumina, 187
aluminum bronze, 117
annealing, 67

brazing, 85

C
    nanotubes, 111
    phases, 21, 103
cathode, 181
cements, 141
coatings, 123, 169
composite particles, 169
composites, 21, 97, 103, 129, 135,
    149, 155
concrete, 135
crystalline materials, 155
cuznal, 31

dislocations, 163

electric properties, 79
extinction, 163

fatigue, 55, 175
fibers, 149
Fourier analysis, 91
fracture, 55
fullerene, 97
functional materials, 187

glassy materials, 155
grain size, 79

high temperature, 49

inconel, 1

liquid bath, 11

mechanical
    milling, 103, 111
    properties, 49, 67
metals, 21, 97
microstructure, 163
modeling, 11, 135

nanoceria, 123
Ni alloys, 1, 49

oxidation, 1

particles, 39
pet, 129
plasma spraying, 169
polymer, 129, 149

random packing, 39

SiO_2 nanoparticles, 141
sol-gel, 141, 181
solidification, 91
spheres, 39
steels, 61, 67, 73, 79
    stainless, 85, 123
stress concentration, 175
synthesis, 181, 187

tempering, 61
texture, 163
thermal evaporation, 31
TiC, 117
TRIP, 73

wear, 61, 117
welding, 73

X-ray diffraction, 163

Printed in the United States
By Bookmasters

Printed in the United States
By Bookmasters